THE URBAN BIRDER

THE URBAN BIRDER

DAVID LINDO

NEW
HOLLAND

First published in 2011 by New Holland Publishers
London • Cape Town • Sydney • Auckland
www.newhollandpublishers.com

Garfield House, 86-88 Edgware Road, London W2 2EA, United Kingdom
80 McKenzie Street, Cape Town, 8001, South Africa
Unit 1, 66 Gibbes Street, Chatswood, NSW 2067, Australia
218 Lake Road, Northcote, Auckland, New Zealand

A CIP catalogue record for this book is available from the British Library.

ISBN 978 1 84773 950 6

Publisher: S...
Editor: Step...
Production:

Reproductio...
Printed and...

Photographs... David Lindo
(pages 35, 9... 79, 84, 134,
140, 144 an... s 36, 53, 97,
114, 116, 1... t (page 11).

CONTENTS

Urban Birding? You've got to be kidding! Cities aren't called the urban jungle for nothing, are they?! Surely they are a no-fly zone for birds…?

Of course, as David Lindo points out in the following pages, nothing could be further from the truth. Britain's cities – and indeed cities all over the world – are at least as good for birds as the surrounding countryside, and sometimes even better. Which when you think about it is just as well, because four out of five Britons, and over half the rest of the world's seven billion inhabitants, live in urban areas – and we all need to get close to nature.

Look closer and it's easy to see why our cities are so good for watching birds and other wildlife. After all, they are packed with green spaces: parks, gardens and areas of woodland or scrub which are ideal for breeding and migrating birds. Almost all of Britain's cities are built close to water: either on a river (the Thames, Tyne and Mersey, to name but three), or by the sea; and the same applies to many cities elsewhere around the globe. Water is always a magnet for birds – whether for wildfowl that winter in huge numbers on city lakes and reservoirs, or for landbirds that simply need a place to drink and bathe.

Cities provide warmth and shelter: thanks to the 'urban heat-island' effect, they are usually a few degrees warmer than the surrounding countryside, and less likely to freeze up in winter. And there's plenty of food, either occurring naturally, or provided by us – accidentally or deliberately.

Which brings me to gardens: that amazing network of 'mini-habitats' that form a patchwork-quilt across our towns and cities. Suburban gardens are, square metre for square metre, more biodiverse than any other habitat in the world, including the Amazon rainforest. They're also a brilliant place to watch birds – whether you are a beginner or an expert, there's always something new to see. So when you look at what's on offer, it gives a whole new meaning to the phrase 'urban jungle'.

If you're going to take a walk in the jungle, you need a guide – and who better than The Urban Birder himself, David Lindo? As David reveals in the final chapter of this book, he and I go back almost two decades to a sunny spring day in London's Richmond Park, where we both successfully twitched an Ortolan Bunting. Since then I have got to know him both as a colleague and a dear friend. We've watched Red Kites soaring over the Metro Centre in Gateshead, some of the capital's last remaining House Sparrows at London Zoo, and on a memorable morning last autumn, absolutely nothing at all flying over Tower 42 in the heart of the City. Well, you can't hit the jackpot every time, can you?

And now David has set down these memories, stories and encounters with urban birds – in Britain and around the world – in this book. Because David and I were both brought up in suburban London, a few years apart, we share experiences of many of the people, places and birds about which he writes so eloquently. But five years ago I left London to live in Somerset, and so now I am The Rural Birder. At least when David comes to visit I can show him the wonders of the English countryside, before he succumbs to the urge to get back home to the Big Smoke.

If you're already a birder, whether urban, suburban or rural, there's plenty to delight you here – tales of encounters with migrants and rarities, epic twitches and memorable days in the field going back almost four decades.

If you're not a birder – or not yet – please don't be put off. For this book should be read by anyone who has ever lived in, travelled through, or visited a city. It contains a vital message: that if we don't notice the wild creatures on our own doorstep, how are we going to look after those farther afield?

For me, the most memorable and moving parts of the book are David's recollections of his early years, being brought up as a second generation black Briton, whose parents had come here from Jamaica in search of a better life. Through David's eyes, ordinary events from a suburban childhood, stories about his Mum and Dad, and schoolboy memories, all combine to create a perceptive portrait – sometimes sad, often very funny, but always fascinating.

You only have to be with David for five minutes to realise that he is one of life's optimists, and a man of great openness, honesty and integrity. He believes passionately in many things, but above all in the power of birds – and especially the birds we encounter in our cities – to enhance, improve and occasionally transform our lives.

Too many of us spend our time staring at the ground. David goes through life staring at the sky, and the infinite promise it may bring; today, tomorrow or in the distant future. He is, quite simply, unique – and so to borrow one of his many analogies to the world of popular culture:

To David Lindo, The Urban Birder – May the Force be with you.

STEPHEN MOSS
Mark, Somerset, 2011

DEDICATIONS

For the men that made a difference.

Collin Flapper, Eric Simms, Peter J Grant, Rupert Hastings and my late father Edgar J Lindo.

ACKNOWLEDGMENTS

Thanks to all the people who have influenced me in their own particular way: Anna Guthrie and The Wildlife Trusts, Barry Hecker, Ceri Levy, Christiano Sossi, Ciska Faulkner, Clare Evans, Clare Lockhart, Dawn Balmer, Deborah Shaw, Delyth Lloyd, Derek Moore, Des McKenzie, Dolly Frankel, Dominic Couzens, Friends of The Scrubs, Gardman, Gary Elton, Hardeep Giani, Helen Babbs, Helen Marsh, Helga Berry, Jamie Oliver, Jay Pond Jones, Jen Hewlett and the Tower 42 Management Team, Jez Blackburn, Joan Scheckel, Kevin Barter, Leela Miller, Liliana dalla Piana, Lucinda Axelsson, Malka Holmes, Natalya Nair, Nick Baker, Peter Alfrey, Richard Chambers, Rick Simpson, RSPB, Sacha Barbato, Sally Cryer, Steve Barron, Stuart Winter, Susie Painter, Swarovski Optik, Tessa Dunlop, Tessa Winship, Tim Hunnable and Nikon, Tim Webb at RSPB London, Vicky Webb, WWT and to everyone else I haven't named.

Special thanks to: Ring Ouzels, Dr Simon Papps, Jo Carlton, Alicia Cieplowska, Tasha Hall, Heidi Birkett, Alan McMahon, Roy Nuttall, Kim Dixon, Stephen Moss, John Charman, Anders Price, Bob Still, Yvette Spencer, Russell Spencer, David Fettes, David Foster, Doug Carnegie, Esther and Bert Higgs, Fiona Barclay, Max Whitby and all at BirdGuides, Jackie Michaelsen, Chris Palmer, Jo Thomas, Kirstine Davidson, Kevin Wilmott, Sheena Harvey and all at *Bird Watching* magazine, Garin Baksa and Opticron, Paul Watts, Scott Crane, Jonnie Scarlett and all at The Quarry, Petria Whelan, Sophie Stafford and *BBC Wildlife* magazine, Penny Hayhurst, Thomas Carty and of course, my Mum x

And thanks to the photographers for all the images supplied: Stephen Daly, Hugh Harrop, Olive May Lindo, René Pop, Russell F Spencer, Sam Twiddy, and Mandy West.

9

Do you love birds?

Do you live in a city or a town?

Do you wake up most mornings wanting to go birding?

Do you feel unfulfilled if you don't get your regular dose of birding?

Do you love it when you are out surrounded by nature, wherever you are?

Does every time you go birding feel more exhilarating than the last?

Do you find yourself furtively searching the skies as you travel around?

Do you ever bore your friends about birds?

Do you get restless during the spring and autumn migration season in anticipation of what might turn up?

Do you find yourself getting up at stupid o'clock to go birding?

Do you believe that anything can turn up at anytime?

Then you too are an urban birder.

I've been an urban birder since I was a small boy and to this day I still wake up with the same excitement for watching birds in city environments as I did then. I set my clock for 5.30am during the summer months and always find myself waking up at 5.20am. Urban birding is the easiest thing in the world to get into and you could literally be city birding in five minutes. All you need to do is look and listen. I led an urban birding walk recently and a few weeks later it was really nice to hear from a woman who was in my group. She sent me an email telling me that she had bought a pair of binoculars and now has the birding bug.

You may not necessarily feel that way after embarking on your first urban birding mission, but I am almost certain that you will be surprised by what you find.

INTRODUCTION

Coming across a stunning bird such as an adult Mediterranean Gull in a totally unexpected location on a city street is what makes urban birding so special.

Recently I went to the funeral of a close family friend in north-west London. It was a beautiful cold, crisp and sunny day with a glorious blue sky in cosmopolitan Harlesden, and two Nigerian ladies caught my eye as they walked down the street in full kaleidoscopic kabas with matching extravagant head-wraps. Whilst looking for a parking space I became aware of some gulls standing on a streetlight opposite a halal butchers a little farther down the busy high road. As I parked I could see through the sun-roof that there were six in total, five were familiar Black-headed Gulls but the end bird looked distinctly different. It was slightly larger, with a stouter reddish bill, longer legs, a dark smudge over its eye and pure white wing-tips. It was a Mediterranean Gull in the middle of urban London, literally yards away from my old secondary school – the place were I had spent years dreaming about when I would ever see this white-winged wonder. Although they occur far more regularly in the UK now, it was still great to see a bird that mainly breeds on isolated lagoons and patches of coastal saltmarsh dotted across Eastern Europe as far east as Turkey, Russia and Ukraine. Potentially, that bird might have been paddling in the Baltic fairly recently, then travelled the breadth of Europe to end up sitting on a London lamp-post eyeing up a butcher's shop.

These kind of events happen every day of every year, everywhere. I learned a long time ago that if you kept your eyes peeled and your mind open then you would begin to see wildlife in urban areas that you just never expected. I've seen a Painted Lady on the top of Tower 42, 600 feet above the city streets; marvelled at Short-eared Owls flapping over urban expanses; and watched Great Crested Grebes flying over a football pitch at Wormwood Scrubs.

There's my Rutger Hauer's final speech in *Bladerunner* moment out of the way. Some might say that any unusual birds in urban areas are chance sightings that should never ordinarily be expected.

Maybe, but I believe that the only reason why they are not reported more frequently is that few people are consciously looking for such occurrences. If you look, then you will see. If you listen, then you will hear. Try it. If you are walking home at night on your familiar route imagine it's the first time that you have ever done that journey. Switch your awareness button on and you might see a fox, or hear or even see an owl. You are nearly guaranteed to hear a Robin singing, and if your stroll is during the autumn migration period, then hearing the shrill 'seeep' sounds of Redwings migrating over our urban centres would not be out of the question.

Seeing unexpected things does not have to equate to unusual or rare birds. Sometimes the usual birds do unusual things. One raucous night at 3am my friends and I were being kicked out from The Electric Private Members Club. Outside the club on a curb on the Portobello Road I saw a Blackbird feeding on a discarded apple from a market stall. It was totally undisturbed by us and I remembered thinking 'shouldn't he be in bed?' Another time, whilst wandering the damp, darkened streets of Ladbroke Grove, looking for night-singing urban Robins for Alan Titchmarsh's series *The Nature Of Britain*, I witnessed a Coot flying overhead at rooftop level. How often do 'unexpected' birds fly over our heads every night?

Most of us are urban birders to some extent, with some people not even realising that when they are feeding and enjoying their garden birds they are urban birding. Others are birders who go on foreign trips and visit some distant hot-spot at the weekend, but consider the thought of seeking birds in a metropolis as not 'proper birding'. What I have noticed quite a lot during the course of my life is that urban birding is often seen as the poor relative, and therefore somehow unimportant. I have visited cities that have really great sites which are sadly underwatched by the local birders.

14

Over 80 per cent of Britain's population lives in urban areas, and around the world the vast majority of mankind now does so. It is predicted that by 2050 at least 75 per cent of the planet's population will be living within urban centres. Unfortunately, the vast majority of those people will be oblivious to nature and unconcerned about conservation. City-living creates a collective feeling that we are somehow above the natural law. We have disregarded our roots. Nature does not apply to us. It is something that occurs on wildlife programmes on TV or, at best, way outside the city limits. Anything that is stumbled across within our urban surroundings is somehow deemed as being there by mistake. The animal encountered may draw a gasp from its urban observer or, more usually, a phone call to the nearest pest controller.

I grew up being continually told that the only wildlife found in cities was pigeons and foxes, a belief that still persists to this day with some folk I meet. Non-birders are always surprised at the number and variety of birds they see when I take them for a walk around the streets of their local area. I always get the chorus of 'Wow, I never knew those birds were here' from astonished novice nature-watchers.

I remember one July day taking out a group of 40 people to Fryent Country Park in north Wembley, on a London Natural History Society walk. The term 'Country Park' was a loose one to describe the place because, despite the rolling meadows and ancient hedgerows, the area was completely encased by suburbia. The people on my walk ranged in age from 4 to 84 and came from all races, colours and creeds. There were not many birds around as most had their heads down, being either involved in chick-rearing or moulting, but the ones we did see resulted in collective oohs and aahs. My group was engrossed when I pointed out a high-flying Hobby, a graceful falcon at the best of times. This one seemed to be aware that it had an audience and proceeded to put on a spectacular aerial display which climaxed in an almighty stoop from a great height. But it was the ordinary birds that really got them going. Seeing a Magpie's iridescent plumage close up was awe-inspiring for them, as was watching a Great Spotted Woodpecker doing its thing on the trunk of a tree. Perhaps the most rewarding moment for me was when I lifted a large log to show the kids the denizens hidden underneath. I expected to find a few woodlice and perhaps a couple of spiders, but instead we uncovered toads, Great Crested Newts and multitudes of millipedes, slugs, beetles and, of course, woodlice and spiders. The children went absolutely mad for it. Out came their mobile phones as they jostled with the adults to take pictures of the creatures on show. It was a full hour before their curiosity expired; yet these were the same people who prior to the walk had not expected to see much because of our urban location.

Conservation starts in cities. If someone living in a tower block can understand the notion of protecting the flowerpot on their window-ledge and get the link that their window-ledge is ultimately connected to the rainforests, then they will understand conservation. It's about education, encouraging kids to discover wildlife, promoting areas for nature within cities and getting people excited about saving their part of the world. It's a big job, but if we can keep the patchwork conservation movement going then one day it will all join up and serious inroads will have been made.

Our global population is growing and our cities are expanding, but maybe it doesn't have to be wholly bad news. Cities are not a substitute for marshlands or pristine forests, and never will be, but we can hope for sensitive development which is sanctioned by educated city authorities that take wildlife into account. Building structures that contain crevices for Swifts and

16

bats, for example, and landscaping that encourages the creation of wildlife-friendly areas to be utilised and enjoyed by both wild creatures and humans alike. It's up to us to make that prospect a widespread and commonplace practice. We need to take the time to uncover the places that are oases for wildlife, and which are often buried in the most unexpected places. It is well worth studying and publicizing these city sites to get them protected as they are sometimes very important as migration stop-off points or as crucial population centres for breeding birds and wintering waterfowl. Being in touch with nature in a city can be a very calming and empowering experience. It can be argued that it is good for our mental health and might make us live longer. At the extreme end of the scale it may contribute to saving the world or at least delaying its destruction by humans.

I would like to take you on my journey of discovery, showing you where my fascination for urban birding came from. But most of all I hope that I might galvanise you to get out and explore your local area. During my journey I have laughed, cried, felt jealousy, maltreated wildlife, spoken up for conservation, been incredibly cheeky, been very scared, seen some amazing places, watched some incredible birds and found my true passion.

Appreciating wildlife can be very satisfying. It's a great way to detox from the stresses of life and to forget all your problems, even if it's only for ten minutes every day. You don't need to have any knowledge and be able to identify everything you hear and see. Birds are everywhere. Just enjoy them for what they are.

I hope that my journey inspires you. And remember, always look up.

18

BABY BIRDER

On a hot summer's afternoon in August 1963 I was taking my first gasps of air in a maternity room at Central Middlesex Hospital in Acton, west London. I was grabbing at what I must have thought was a pair of binoculars, only to find that it was the doctor's stethoscope. To be honest, that would be just the type of story I would love to be telling my grandchildren as I bounced them on my knee as an elderly man, harping on about days gone by. I would probably be responding to the question posed by one of my imaginary grand-infants: "Why did you call yourself The Urban Birder, Grandad?"

A lot of people think I woke up one morning and decided to call myself The Urban Birder from that day onwards. Well, they're kind of right in one sense. But the story stems much further back than that, and it's the experiences and adventures I had along the way that have shaped and guided me into what I am trying to achieve today. We will come to define what The Urban Birder really is in due course, but first I'm taking you back to those initial few bird-filled years after my birth. I would like to be

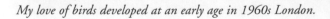

My love of birds developed at an early age in 1960s London.

able to say that there was a seminal ornithological moment that dictated my interest in birds and nature. I'm thinking along the lines of a Robin landing on my mother's pregnant belly or a migrant Ring Ouzel flying into the delivery room and landing calmly on the midwife's shoulder to monitor proceedings. That last notion is not completely out of court, as I was born on 22nd August, which is right in the mix when it comes to autumn migration. Time enough for an early migrant Mountain Thrush, to give the Ring Ouzel one of its old vernacular names. But it didn't happen. Even so, a few years later this handsome and elusive thrush was to become my favourite bird.

21

My mother told me that as a child I was pretty untrusting of anyone that I didn't already know. If you tried to pick me up it would invariably result in plenty of wailing and the thrashing of limbs, as a dentist once testified after receiving a kick in the face when he tried to place me in the chair. I also caught a very bad case of mumps as a tot. In the '60s mumps and chickenpox were feared diseases, especially for infants. Apparently, when I had it my face was extremely swollen and I was in a very poorly state. I certainly had it bad. So bad, in fact, that it affected the aural nerve in my left ear, practically rendering me deaf in that ear for the rest of my life. To this day I often have trouble picking out and locating bird calls, due to my lack of aural spatial awareness.

My parents were immigrants from Jamaica, the second or third wave after the *Windrush*. My mum Jean arrived in 1960 and my dad Edgar a year earlier. They knew each other from their homeland and married soon after they hooked up in London. They had come from an entirely different culture to live in an environment that had different weather systems and was inhabited by a different race of people, a few of whom were less than delighted to have newly arrived blacks around. Not that my parents ever complained about racial prejudice.

My mum and dad worked hard to put food on the table and to keep a roof over all our heads. My dad was a very proud man who believed in the Jamaican and decidedly masculine tradition of being the firm hand of discipline in the house: a situation that by the time I was a teenager was to finally drive us apart. Conversely, my mum is a quiet, shy woman who ordinarily wouldn't say boo to a goose. Neither had any interest in birds, or in any wildlife for that matter. They used to speak nostalgically about Jamaica, and when I asked about the wildlife they would sometimes make casual references to 'John Crows' (Turkey Vultures) and 'Doctor Birds' (Red-billed Streamertail – an endemic species of hummingbird that is also the island's national bird). They were not paying homage to Jamaican natural history, but instead issuing a nod to a romantic and idyllic past.

22

Doctor Bird's label – note the streamertail image on the left-hand side.

We used to have a gramophone cabinet in the living room – the classic West Indian furniture of choice, along with the plastic covering over the best sofa. My dad loved his ska, reggae, Nat King Cole and country and western. He used to spend hours playing his 45s and there was one record that caught my attention, by Alton Ellis and The Flames. The track was called *I've Got A Date*, and if you see me and ask me nicely I might whistle the tune for you. It certainly was not a classic, having its day in 1967 or thereabouts, then submerging into obscurity. But the thing that attracted me to it was its label. It had a green-and-yellow background with a stylised bird with a long beak and tail streamers slapped to the left of the spindle hole. A Doctor Bird.

Quizzing my dad about whether he had seen one before and what it looked like resulted in frustration for me. The poor guy didn't take my questioning seriously, nor had he really looked at a streamertail – or at any other bird for that matter, unless it was cooked and served up on the table. I was a kid in the '60s, the Internet hadn't been invented, and books on foreign birds were not even a consideration. I had no clue what a Red-billed Streamertail looked like, but on reflection I don't blame him for his inability to satisfy my curiosity, as he was not to know then just how strongly my fledgling interest in nature was to grow.

A naturally happy child with a sunny disposition, I was more than capable of being gregarious, although this was sometimes tempered by my propensity for withdrawing into a shell and being more of a loner. This latter trait was borne out of the large amount of time I used to spend by myself. The pursuit of observing wildlife certainly brought this out. I guess I have always been an explorer. I was inquisitive and relished walking down a path to see where it would take me; and I certainly loved forging new ones as I explored the nooks and crannies of my neighbourhood.

Roundly ignoring my toy yacht as I scan the sky for passing raptors.

25

My earliest memory of exploration was as a three-year-old, when my parents and I lived in shared rented accommodation in Willesden, north-west London. It was a simple flat with no bathroom. If you wanted to have a wash it meant a trip to the kitchen to use the portable communal tin bath. One day my mum took me to a local West Indian house party. In those days it was the West Indian tradition to start a party during the day so that children could attend, and then by nightfall the kids would have gone and the hardcore action could begin. I'm not sure what the occasion was, although it must have been a Sunday as we were all dressed up. I looked very natty in a fetching pair of shorts. Someone gave me a toy yacht at the party which occupied me for a brief period until I got bored. I had this incredible urge to go outside and escape the boredom for some adventure. I slipped outside into a grey, overcast day. No one had noticed me leave as they were all busy singing, dancing and drinking, and soon I was strolling the streets of Willesden, yacht in hand, eyes agog.

Meanwhile, back at the gathering, pandemonium had broken out. My mum was screaming with despair, the police were called and a full search-party was quickly dispatched around Willesden. My mum recalls running through the streets stopping everyone she met to ask if they had seen a wandering small black boy in shorts with a yacht. My mum eventually found me standing in front of a small churchyard, apparently staring at some birds that had held my attention.

A year later I experienced the first of my three wildlife epiphanies. This initial eureka moment was borne out of yet more boredom. I was not in the middle of a nature reserve; instead it happened via the pages of a coffee table book featuring action pictures of African wildlife. I discovered this

26

book whilst on a boring visit with my parents to their friends' house in Hornsey, north London. Laurel and Seaton had known my parents since back in the day in Jamaica. Their house had the typical West Indian styling of the time with classic plastic-covered furniture and white crochet doilies on the backs of the sofas. Laurel and Seaton were a bit older than my parents and seemed ancient to me even then, though I really liked them. But I still found visiting their house a chore until I stumbled across a wildlife book they owned. It probably was an average-sized book, but it seemed massive to me and soon I became deeply engrossed by the images of scattering zebra and wildebeest within its pages. Sitting with the book by their coffee table I would transport myself to be galloping amongst the herds. I studied every detail of their faces, coats and movements as everything around me blurred into the distance. My interest multiplied from that moment on, and subsequent visits to Laurel and Seaton's became a great pleasure.

My interest in natural history may have been inexplicable, but it just kept growing. In 1968 my mum and dad bought a semi-detached house in Wembley, in the then suburban district of the now invisible county of Middlesex. It's now quite urban and widely considered to be part of north-west London. I remember us pitching up to the house to view it one evening to find it filled with boy scouts. The owners were hosting the local scout meeting, but for me it was the first time that I had ever

27

seen anything like it. I really liked the house: not so much because I had my own room, a boxroom overlooking the street, but because we had a long garden. Heaven! The house was also situated on Wyld Way. Was that another thinly disguised omen? When we moved in my mum was pregnant with my sister, and it was also at this time that I started primary school. I was lucky because my school, Oakington Manor, was literally around the block from the house. I was also fortunate that there was a small wood at the end of the playing fields, within the school grounds. Many an afternoon was spent exploring this oasis in the middle of urbanity.

28

Meanwhile, back in my new home I was really getting settled in. Our house was part of a block of 1930s semi-detached housing, encircling loads of narrow back gardens. All the houses were identical in their layout with a kitchen, larder, dining room and sitting room downstairs, and three bedrooms (including two doubles), a toilet and a bathroom upstairs. We lived in a quiet, predominantly white middle-class area, a stone's throw from Wembley Stadium. The impressive and iconic twin towers (now sadly demolished) were clearly visible from my upstairs bedroom window. I could even hear the baying football crowds during cup finals and internationals and Elton John once serenaded me in my bedroom from the stadium. During the summer, the calming sounds of endless lawnmowers, dogs, children shouting, adults barking orders and, above all, birdsong filled the air. In the winter the dogs still barked and sparrows chirped. That was the soundtrack which – especially the mowers – will forever remind me of my younger years in Wembley.

It was while I was exploring in the school woods that I first came across the myriad invertebrates to be found on the woodland floor. Woodlice, caterpillars and spiders of all sizes became my friends. I intently studied the movement of the millipedes I discovered and any beetle that crossed my path was quickly collected. I would bring these specimens into class in the vain hope that they might be recognised by a teacher.

As a small child I just assumed that anyone who towered over me in height didn't get that high for no good reason. Like most kids I held the view that all grown-ups knew everything about everything. I was wrong. Not one of the teachers could give me the courtesy of a considered answer to the identity of the creatures I had unearthed. I simply was not happy with the bog standard 'it's a creepy-crawly, David' response. It just wasn't good enough. I mean, I remember being in an art class and seeing one of my schoolmates draw a picture of a bird. Nothing wrong with that you might say. Well, there wouldn't have been anything wrong were it not for the fact that the bird had four stick legs. 'Miss!' I shouted, 'Birds don't have four legs!' I remember being horrified when the young teacher told me that they did. For a few moments I began to seriously doubt myself.

My growing interest in natural history, and in particular birds, soon became apparent to my school friends and I was quickly re-christened 'Bird Brain'. The majority of the kids were English, but there was a small cohort of first generation West Indian, Asian and Irish kids. Racism was rife, although at the time it seemed natural, as if it were simply a fact of life. It was 'nig-nog this' and 'blackie that'. I knew at the time that they were simply repeating words and phrases they had either heard from their parents or learnt from watching popular sitcoms like *Love Thy Neighbour* and *Till Death Us Do Part* with Alf Garnett. The former programme was a particular favourite in my household, along with *It's A Knockout*. My dad and I used to cry with laughter watching these shows, as we considered them genuinely funny. So when I heard someone calling me a golliwog in the playground I associated it with humour.

Once a kid called me a sambo, expecting me to burst into tears or run off. I replied 'Stan Bowles? Oh thank you!' Stan Bowles was a gifted footballer who played for Queens Park Rangers. Now annoyed, the kid spurted 'No! Sambo! SAMBO!'. To which I replied 'Cheers! Stan Bowles! Gee!' and that was the end of that confrontation. Sometimes I would answer back racist taunts by calling people 'wormlips' or 'snowflake', a name that I nicked from *Love Thy Neighbour*. Both names seemed silly to me, even at the time. Although I tried to treat racism with a pinch of salt it did unsettle me at times, especially when I began to realise the true ugly malice that it occasionally concealed. That became particularly evident from my teenage years onwards. Eventually I began to become very aware of my skin colour, to the point that I sometimes felt inferior and cautious when I met white people I didn't know.

The other big things at Oakington Manor were Corgi car collecting and racing, conker fights and marble swapping. All were a big deal in the playground but it was the trade in ladybirds that took things to another level. We used to go around with small sweet tins collecting the insects from the playing field and the school wood. Some of us used to have tins filled with scores of these creatures, but I was the king 'ladybirder'. I may not have ruled when it came to Corgi car racing (although I wasn't too bad when it came to possessing killer conkers) but I used my knowledge of the school wood to eke out bountiful ladybird hunting grounds, and I would often spread the search to my back garden.

The ten-spot and the other regular red examples were the ones we based our economy on. It was the yellow ones of any spotted denomination that were deemed as valuable as gold dust. Just one of those babies was worth at least ten of the red ones. We used to spend all playtime searching for, and then swapping, them. This pastime continued until the summer of 1973, when there was a huge influx from the continent. The playground market was literally flooded. There were

31

thousands everywhere. Once, whilst trying to fill a container full of the beasts, one of them bit me. I was in shock. The very insect that was the symbol of placidity had in one sharp nip shattered my comfy warm feeling for them. My ladybird-collecting ended precisely at that moment.

At home I was conducting regular invertebrate collection expeditions into my garden, which a few years later would be fuelled by the exploits in the Gerald Durrell books I read so avidly. I noticed that the deep crevices of our two garden sheds, particularly the rickety one at the end of the garden, housed countless numbers of weird leggy spiders, silverfish and earwigs. All these creatures existed in the house as well, but there was something about finding them in the wilder, undiscovered corners of the garden. So in 1969 I created my first nature reserve. It encompassed several large tracts of the garden including a postage stamp-sized patch of rough grassland halfway down, the grassland under the apple tree and the large barrel by the side of the shed nearest to the house that held the drainage water. The barrel was used as a spawning ground for mosquitoes. I used to watch the wiggly nymphs for ages, not realising that they would morph into something quite sinister.

I also stalked the garden searching for bumblebees. I found them fascinating. How could something so fat fly? I used to collect them with an empty glass Coca-Cola bottle, picking them off the flowers they were feeding on and holding them in my hands before stuffing them down the neck of the bottle. Eventually, when the bottle was filled I would empty them all into the barrel of water. The reason for my abject cruelty? It was all in the name of science. I wanted to see if they could swim. Of course, they couldn't; they all drowned. I learnt a valuable lesson. Bees and water do not mix. I felt a bit bad so I begged my dad to fix the lid on the barrel to prevent any more bees from accidentally drowning in the open water.

I was warden, curator, architect and chief cat-scarer of my managed natural domain. Butterflies were not safe from my gaze either. A cheap plastic net on a pole bought from the local newsagent became my chief butterfly-catching implement. The net was originally designed for scooping fish out of ponds, so the netting was not the fine mesh that, with a graceful swish, would have caught the delicate beasts, causing them no harm. Instead, as I whooshed and swiped after an unfortunate Cabbage White it would end up at very best with a vast amount of the scales on its wings missing, or worse still with no wings at all and, to all intents and purposes, dead. Thankfully, I learned that perhaps I wasn't employing the best method of catching them so I soon gave up the butterfly-collecting.

Meanwhile the small area of thistles and grasses I had cultivated, directly influenced by the grassy areas I had found at primary school, had inadvertently become a haven for a variety of wildlife including froghoppers. Their froth-covered nymphs interested me, and I didn't discover that they would grow up to be froghoppers until years later. With a sturdy piece of grass stem I would gently scrape away the spittle to reveal an ugly creature gripping onto the stem of a plant. I also created a small rockery near the back door where I re-housed all the creepy crawlies I could find, sometimes with unintentionally devastating consequences. Hordes of woodlice, snails, ants and too many spiders in close proximity were definitely not a good idea!

As far as I can remember, birds entered my life during the summer of 1969, a few months before the birth of my sister. It was my second epiphany. On the day of my sister's birth my dad took me to the Central Middlesex Hospital to meet my new sibling. Instead of rushing in with him to see her I stayed outside to count the House Sparrows on the hospital's green. I was completely taken by the group as they flew from the ground to nearby bushes, only to return once they felt safe. How did they fly? What were they doing? How many of them were there? I didn't get a chance to work out the answers to those questions because my dad had come back out to drag me in to see my sister.

Susan, as she was named, was given the large back bedroom as a nursery, and every day after school my mum would leave to work nights at the McVitie's factory in Harlesden, a couple

Counting House Sparrows – more interesting than the birth of my sister?

of miles away. I would have to babysit my sister until my dad came home from his welding job in Neasden. Playing with Susan in her cot, or staring out of the back window over all the gardens, filled the two-hour gap. It was then that I really started noticing birds en masse. At first I thought that all the

35

36

'Spoon Wings' were regularly sighted over our Wembley garden, even before I knew that their correct name was Lapwing. I was only six years old!

birds I saw flying around were being puppeteered by God, and I could have sworn that I actually saw strings attached to them. Thankfully, that belief was short-lived.

I noticed birds coming into the garden to the soundtrack provided by John Peel on Radio One. The Move, Paul

McCartney and Captain Beefheart provided some of the musical backdrop whilst I watched the birdlife outside. At first they were unidentifiable but yet at the same time somehow distinctive. Some were big, some small; some were dun-coloured, others had a distinct amount of colour in their plumage. I didn't yet have a field guide, so I had to improvise. I found that 'mummy' birds and 'daddy' birds (Starlings and Blackbirds respectively), 'baby birds' and 'uncle birds' (House Sparrows and Carrion Crows), were all common. I would watch them for hours, fascinated by the fact that they were so easily seen once you started looking. From my lofty perch I watched Woodpigeons performing their characteristic undulating display flights and christened them 'Jack-in-the-Boxes'; while the blunt, rounded wings of the Lapwings regularly passing overhead reminded me of spoons, so they were named 'Spoon Wings'. I was only six, and not very tall, so to conduct my early observations I peered out of the back window by balancing on the pipes below the windowsill that ran from the airing cupboard to my right to the bathroom, the next room behind the wall to my left. I soon learnt to ram the dressing table behind me for a more comfortable prop. I had become a birder, with my own hide, and I didn't even know it.

Eventually, I got myself down to the local library and obtained a couple of field guides on loan. One was the

37

dinky-sized *Observer's Book of Birds* replete with its tiny images and the other was a Ladybird book on British birds. Both gave me the impression that birds were best looked for in the countryside. But I was stuck in Wembley with no one to take me anywhere so I had to learn that the best time to observe the action was first thing in the morning, before the neighbours came out to mow their lawns. The drawback with pre-school, early-morning, bedroom-window birding was that I was sometimes the unwitting observer of showering women in the bathrooms opposite me. The potential for being labelled as a peeping tom was enormous, but fortunately very few people ever noticed me and, anyway, I always tried to avert my gaze!

With the help of those books I began to put names to the various species I had been seeing, and by the time I was eight I had begun to keep a list of the birds I recognised in my garden. That list included a few really dodgy sightings, such as three 'Peregrines' flying over (undoubtedly Kestrels) and a 'Pied Flycatcher' that had to have been a Pied Wagtail – however, to give me some credit, I did only have brief views.

Officially, my first properly recognised Kestrel was the one I saw over my school woods on a summer morning in 1970. It landed within the wood and, totally excited, I rushed in to tell my headmaster, Mr Flanaghan. 'Sir, I've just seen a Kestrel!' I blurted out. His response was less than enthusing. He looked down at me, and uttered the immortal words:

'David, you haven't seen a Kestrel. You don't get Kestrels in cities.' I knew that I had seen a Kestrel, and I knew that he was wrong. Mr Flanaghan was a big man and quite old in my young eyes – all the things that symbolised authority and knowledge. But he was wrong. From that moment I subconsciously realised that there was more to wildlife in cities than I was being led to believe. I was on a mission.

A few weeks later, during playtime, I found a fledgling House Sparrow sitting on a branch at the edge of the woodland. It sat there all alone, chirping for its unseen mother. I was enthralled. I had never approached a wild bird that closely before and soon I was touching the confused youngster. It was so delicate and was trembling. For the previous few months I had become obsessed with trying to capture birds, particularly sparrows as they were so common. I wanted to stick them in a cage, ostensibly to study their habits, but in reality simply to own a pet bird.

My plan was quite simple: to lure the birds close to me in my back garden by acting like a bird myself. Genius! This cunning plan involved sticking bits of old newspapers on myself to replicate feathers and squatting near the back door surrounded by bits of white bread with my arms folded behind my back like a set of wings. Oh, did I forget to mention that I was wearing an emerald green, cone-shaped Christmas party hat over my face as a makeshift beak? I thought that I looked like a Woodpigeon, and that the birds

39

would accept me as one of their own. Once a bird got close enough, my plan was to grab it and quickly deposit it into a waiting rusty cage I'd secreted in the nearby shed. Needless to say, I did not get within a mile of any passing bird. They must have looked down at me from their lofty perches and thought 'what is that idiot doing?' To think of how I must have looked still fills me with acute embarrassment.

Back at my school woods my newly-found fledgling sparrow, now christened Fudge, was getting first-class treatment and being fed the finest scraps of bread from the school canteen. I had enlisted a few of my mates to help gather caterpillars and other grubs to feed him. I was so happy to have at last found my own pet bird. The lunchtime play period drew to a close so I made sure that Fudge had enough to eat and I ran off to my lessons. My thinking was that I would return after school to collect my little sparrow and take him home. Of course, when I returned Fudge was nowhere to be seen. I searched everywhere: in the bushes nearby, in the branches overhead, but I could find no sign of my little Fudge. Dejected by the thought that he may have been reunited with his natural mother or, worse still, fallen prey to some unknown assailant, I sloped off home.

As my love for birds grew, I commandeered the apple tree from which I hung all manner of dodgily constructed home-made bird tables fashioned from fruit crates. The first table I created was impaled on a post that I had staked into the lawn without getting planning permission from my parents. It failed, because not only did the food fall through the gaps in the crate, but a sudden gust of wind sent the construction crashing to the ground.

Cheap plastic nut bags bought at the local corner shop with hard-earned pocket money were a better bet for attracting birds into my garden. Back then the bag itself was made with cheap harsh plastic and was very different from the ultra-fine material of today. However, I had long realised that putting out food was a brilliant way of getting to see birds close up from the comfort of the back bedroom window. I was rewarded by great views of Blue and Great Tits on the nut bags, and Blackbirds trying to balance themselves on my decidedly unstable feeding station.

It was also around this time I first became aware of the Young Ornithologists' Club, which was better known as the YOC – the junior arm of the Royal Society for the Protection of Birds. I talked my mum into paying for my membership and before long I was avidly reading *Bird Life*, the membership magazine that periodically came through the letterbox. I scrutinised every single black-and-white image and read every single word several times over. It was whilst reading the advertisements that I came across two things. The first was a contact advert from another similarly-aged local boy, who wanted to get in touch with another junior enthusiast for some birding. On the next page was an advertisement for a must-have gadget that every kid depicted in the publication had – a pair of binoculars. That was what I had been missing for all my short birdwatching life. I had to get a pair.

42

ALWAYS AND FOREVER

My eighth year on earth was a major turning point in my life, a pivotal time that profoundly shaped my future interest in birds and wildlife. I made the transition from just being fascinated to being totally possessed by a thirst for birding knowledge. It was my baptism. My poor mum. I mean that quite literally as she wasn't at all flush with cash. When I started pestering her for a pair of binoculars as an eight-year-old she was still doing the night shift working in the McVitie's factory, churning out those biscuits to help make ends meet. She would return home after her shift laden with packets of the stuff. Custard creams, digestives, rich teas and cream crackers – all good fodder for the makeshift bird table.

Meanwhile, I had my heart set on a pair of 10x50 binoculars I had noticed in Dixons on Wembley High Street. They were a bulky pair that would not have been out of place on a naval captain's bridge. The £14.99 price tag may not sound like much in today's money, but it was a major deal then. After a few weeks of lobbying my mum eventually bought them on hire purchase, a payment type that meant little to me. All I was concerned with was getting my little hands on those giant optics and gazing down a sun-emblazoned Wembley High Street through my newly acquired bins. I was truly ready to start my long birdwatching apprenticeship.

The world outside my back window took on a whole new dimension, but viewing the birds was not as easy as I had first imagined. Wielding heavy, oversized bins was difficult, as I had trouble trying to find the birds that I could see easily with my naked eye, so I had to learn hand to eye co-ordination. Having no mentor and no reference points I had to improvise. That meant many hours standing in the garden training my binoculars on relatively slow-moving passing aircraft, perfecting the move like a sharp-shooter spinning on my heels to catch the plane and focus on it at the same time. After a few days I moved on to birds, and soon I was catching and tracking the Starlings that whizzed overhead, as well as the occasional Woodpigeon.

When things got quiet in terms of animate and inanimate aerial traffic I used to indulge in perhaps the most embarrassing activity I could possibly carry out in broad daylight. I would stand in my garden and at the top of my voice start shrieking outlandishly bad impersonations of Golden Eagle and Great Northern Diver: two species of birds that would never be found anywhere near my garden in north-west London. Both have classic cries: the eagle has the archetypal shriek and the diver a hauntingly chilling wail which has been sampled by many a Hammer House of Horrors flick and by cool house music producers. At the time I thought my bird calls were note-perfect, and that if any of those birds were in the vicinity they would fly over my head to investigate. I guess this is where my eternal optimism first reared its head. What must the neighbours have thought?

On one of my regular fact-finding trips to the local library I discovered a new bird book: *Birds of Britain and Europe with North Africa and the Middle East* by Heinzel, Fitter and Parslow. Flicking through its pages and seeing the multitude of species, many of which I had never even heard of before, was simply mind-blowing. It was as if I had found the sacred scripture. For the first time in my short life I had all the species likely to be seen in what birders call the 'Western Palearctic', complete with distribution maps and illustrations of the different plumages. The guide to dog breeds of the world that had been my previous fixation was swiftly traded in, and I began scouring the pages of my new birding bible.

Needless to say, that copy never made it back to the library. Instead I took it everywhere, reading it religiously, even at school. My already established 'Bird Brain' moniker was truly reinforced when my school friends saw how engrossed I was with this book. It was not long before I had become a walking compendium of European birds. I had memorised every single species featured in the book by sight, every scientific name, and even the length of each bird.

45

My favourite species was the Ring Ouzel, probably because it seemed so familiar, looking like a piebald Blackbird, but at the same time it was totally alien. I thought that I would never see one. When would I ever get to Yorkshire?

It was one thing knowing all the species in the book, but now I needed to see them in real life. But how was I going to see a mountain-loving Ptarmigan in Wembley, or hear the call of the Fulvous Babbler – a bird that normally lurks in North Africa – along the High Road? There was only one way I could see exotic birds without venturing far from my home – I would have to go to London Zoo. My thinking was that if I could see what they actually looked like in the flesh then when it came to seeing them in the wild it would be like renewing an old acquaintance.

So one weekend I went off to London Zoo with a couple of my mates. The zoo adjoined Regents Park and, as we had no intention of paying to get in, after a bit of scouting we discovered that there was a hole in the perimeter fence, funnily enough near the Lion's cage. It was just big enough for us to squeeze through. Once in I ended up going off to look at the birds as my mates ran amok, more excited by the fact that they had just bunked in. I remember thinking how large the forlorn-looking Oystercatchers were and how queer-looking the Stone Curlews seemed. Overall, the birds I saw were more charismatic than their images in the book. I did the zoo thing a few more times, once with the school and once rather audaciously by climbing over a fence in full view of the paying public. Thank God that CCTV had not been invented yet. I am sure that the zoo has improved upon its monitoring of unscrupulous kids trying to gain entry for free; and besides, may I stress that since those unruly days I am now a totally reformed character!

My best friend at primary school was Colin Gray. He was a black kid who lived on the notorious Chalkhill Estate, about a mile to the north of my home and quite close to Brent Reservoir. This was a site I never even knew existed at the time, but in future years it would become my first proper local patch. Colin lived in a high-density tower block housing estate that was beginning to get a bad reputation for crime; in fact it became so much of a no-go area it was eventually demolished in the 1980s.

Colin was intrigued by my passion for natural history and took more than a passing interest. He would sometimes accompany me on my expeditions into the school woods, and even went as far as getting a pair of binoculars, but we never really went birding together due to his reluctance to be seen out watching birds by his mates. As some of us may have experienced, kids can be very cruel to each other. It is also a very impressionable period in life. The things that happen to you during this vulnerable time can make, break and shape your later life.

As well as being derided by a lot of the kids at school for

47

being interested in birds, I was especially ridiculed by some of my fellow black brethren. They somehow felt that I had let the side down. This early experience left me very wary of exposing my interest to other black people well into my adulthood. By and large the black kids at my primary school all came from roughly the same background as me: first generation West Indian immigrants. Most of them just could not get their heads around my predilection for birds. They simply could not understand it. Some of the more militant and threatening boys accused me of being 'white', 'English' and a 'coconut' – black on the outside and white on the inside. These were accusations that really hurt, causing me to consciously rein in my interest and become more cautious to avoid being singled out. As for the black girls, they would not give me a second glance, which wasn't a problem as I had no interest in them anyway.

When cornered I used to defiantly tell my black anti-black birder antagonists that at least I was doing something different and I questioned what they were doing that was different. This did not go down well, especially when they realised that they really were just being sheep. Despite this, there were a few kids that embraced my developing passion. Not surprisingly, they were largely white kids, several of whom were regular visitors to that expansive and mythical wilderness that I so yearned to visit – the countryside. Birding was referred to as bird spotting in those days and was deemed

49

Sometimes being a young birder in the 1970s was far from an easy business.

to be the preserve of the rurally-based white, tweed-wearing, walking-stick-brandishing middle classes. So what was a young working class black kid doing getting involved?

When my parents' generation first arrived on these shores from the West Indies back in the '50s and '60s, most came from rural backgrounds and had strong relationships with their native flora and fauna. But once they arrived in Britain's cities their survival priorities immediately kicked in. Any leisure time was spent bonding with mates over rum and raucous domino games, whilst dancing to the latest ska tune. Add to that scenario the fear and trepidation that many ethnic immigrants felt when they imagined venturing outside the city and into the English countryside and you start to generate nervousness. Having a house party or chilling with mates during any downtime were the only options – forget about a stroll in the countryside to take in the rural chorus of Yellowhammers and Skylarks.

I really feel that in those days newly-arrived immigrants were convinced that if they did end up in a quaint country village they would have been gawped at, racially abused and chased out of town by the Alf Garnett-type locals. These beliefs were fuelled by the images of British life as portrayed by the media at the time. You can see why they sought safety in numbers by grouping together within selected areas of the cities. Besides, why get yourself into a fish out of water situation when you can stay within your familiar territory?

My first birding buddy was a kid I contacted through the classified ads in the YOC's *Bird Life* magazine. I have completely forgotten his name now, but he was a little blond kid of a similar age to me. After his mum had rung my mum to do all the checks and I'd had the day pass granted by my dad, I went to meet my new birding buddy and his mum in the posher suburbs of Hatch End, a few miles to the north. I had a great afternoon with my stolen bird guide and new mate watching woodland birds in greater abundance than I had ever seen before. It helped that it was a beautiful summer's day, as everything seemed more colourful, vibrant and magical. Blue and Great Tits proliferated and I saw my first Coal Tits and perky Nuthatches in the woods near his house. The abundant Collared Doves intrigued me – I just didn't see them near my home.

To close a memorable day I 'saw' an Azure Tit perched on a suburban aerial. This was the problem of using a field guide that included many species that are not found in Britain. Azure Tits are like long-tailed, hoary-looking Blue Tits, and have never been recorded further west of their Russian range than Poland. For me, being an eternal optimist literally meant that anything could turn up anywhere and at anytime. As I gained more experience I would learn to temper this wishful thinking.

The nameless boy and I met a couple more times, but we never really gelled and soon drifted apart. But that didn't matter because I was on a mission to see as many birds as possible, and was lost in a world of books that I pored over with gusto. I read Gerald Durrell's books from cover-to-cover, imagining myself growing up and travelling the world

51

collecting rare animals for my own private zoo. Gerald Durrell was a great naturalist, born in the 1920s, who went on to set up his own zoo on Jersey. He held the distinction of being my first naturalist hero. His tales of stalking animals in far-flung places served to inspire my young mind and made me realise that what I was getting into was much more than simply watching birds. It made me pick up an exercise book and begin to note down the birds I had seen in my garden. I recorded all sorts of stuff, ranging from the expected Robins and Blackbirds to some rather fanciful ornithological delights like the previously mentioned flyover Pied Flycatcher that should really have been hanging out in a deciduous woodland in Wales and a selection of oddly-marked, hitherto undescribed species that were surely new to science.

On my eighth birthday my mum gave me *Birds of the World* by the late Oliver J Austin Jr, with illustrations by one of America's best known 20th-century bird artists, Arthur Singer. Published in the 1960s, it was a large-format book filled with images of exotic birds spanning all the known avian families around the globe. I spent many happy hours poring through that tome, resulting in me filling even more notebooks with handwritten lists of birds that were in danger of extinction. I began to research and write a checklist of the world's birds, a job which eventually developed into an ongoing lifelong task, and evolved from paper to electronic format.

A spread from my first book 'Thrushes and their Enemies'.

53

Birds of the World also inspired me to write my first book at the tender age of nine, entitled *Thrushes and their Enemies*. This may sound amazing, but it was a work that almost completely plagiarised *Birds of the World*. I had an interesting choice of supposed thrush 'enemies' that included Golden Eagle. Similarly, I lumped species together that I deemed to be thrushes, including unrelated ones such as orioles. I illustrated the book by totally ripping off Arthur Singer's images, colouring in my attempts with felt-tip pens. Nonetheless, I was pretty pleased with my work, and when I took it into school to show some of my teachers I was given a straight A+ even though it had nothing to do with any schoolwork.

My parents were beginning to realise that this was no passing interest that I was developing. They didn't really understand it, and must have exchanged quizzical glances, but my mum especially was very supportive. My dad never really openly commented much about my love for birds, but he didn't discourage it. One day he came home from work and announced that an English work colleague of his who had an interest in birds was coming to take me out over the weekend. The news was like manna from heaven and I excitedly started to count down the minutes until our day out.

My dad's friend eventually rolled up in his Morris Minor the following Saturday morning. When he stepped out of the car I noted that he was an older man who looked like a classic union rep, with voluminous sideburns and moustache. Well, it was the early '70s. We left after he'd had a cup of tea with my dad and went on what seemed like a mega journey. I could barely contain my excitement. He took me to the now-defunct Radlett Gravel Pits in Hertfordshire, some 20 miles from my house – crucially for me, in the fabled countryside. We walked into the site and immediately saw Lapwings, my erstwhile 'Spoon-wings', indulging in their crazy, floppy display flights. My dad's friend had certainly been around for a while because he actually referred to them by their old name, 'Peewit'. It was the first time that I had ever seen their display and heard their distinctive 'peewit' calls. My previous experience of this species was the flocks that frequently passed over my garden, especially during June and July. Those were the days prior to the species' dramatic decrease in Britain. However, they were a common-enough sight for me in my north London neighbourhood that I greeted the flocks I saw every year with nonchalance.

Walking through some scrub we flushed some birds that he simply called 'partridges'. I did wonder whether they were Grey or Red-legged, but I wasn't too concerned because either way I had never seen a partridge of any kind before. We sat quietly in a clearing and listened to birdsong. He pointed out singing Willow Warblers, Chiffchaffs and Blackcaps, as he taught me about waiting for birds to show themselves. As an active egger, he knew all about nest-finding. His ability was unerring. That afternoon we discovered a Blackcap nest after watching the parents repeatedly returning to it. In between their visits he took me to the nest and gently pushed aside the foliage to briefly show me the chicks. I had never noticed nesting behaviour before so I was transfixed.

55

In hindsight I was very grateful that he concentrated on teaching me fieldcraft and didn't tempt me into collecting eggs because, being an impressionable young boy, I certainly would have done. Of course nowadays the thought of a boy being taken into the woods by an older, unrelated man with bushy sideburns and moustache would be unthinkable, given the politically correct world that we inhabit.

My third and final epiphany occurred when I was around ten. By then I was a complete bona fide bird nut, so how could I need any more convincing? One Sunday afternoon I was sitting on the couch in front of the television playing with my sister when an animated film came on. It was called *The Last Of The Curlews* and it immediately caught my attention. It was about the life of the last Eskimo Curlew, following its migration from its arctic breeding grounds, across the Americas, to Argentina and back. The curlew, a male, connects up with a bunch of American Golden Plovers on migration and becomes their leader, but all the while he's looking out for the non-existent flocks of his own species. He has frequent flashbacks to the days when there were plenty of Eskimo Curlews, before they were shot out of existence. Finally, he hooks up with a lone female in his winter quarters and they proceed to head north all loved-up.

56

However, tragedy strikes when a trigger-happy farmer in the Midwest shoots the female as the couple are roosting in a field. They both take off and head onwards, but the female soon expires on a woodland floor. The mourning male sticks by her side the whole night until he is disturbed in the morning by a little boy on a hunting trip with his father. This boy had unwittingly seen the lone male head south earlier in the spring. The story ends with the male continuing his lonely journey north.

This film contained a great conservation message, and I remember being very upset by the wanton wholesale killing of these birds. I turned to my sister with tears streaming down my face and begged her to tell me why they were killing the curlews. She could only look at me blankly before continuing to fumble with her teddy. She was only two. I watched it again on YouTube the other day for the first time since I was a kid and despite the very syrupy moments and the sometimes cringeworthy anthropomorphic bits it still pulled at my heartstrings.

The Last Of The Curlews had a dramatic and everlasting effect on me. From initially seeing birds as creatures that just happened to fly about in my garden, I realised how fragile their existence really was. It made me angry at the futility of hunting for the sake of killing. The film also fired up my desire to learn more about conservation, locally, nationally and internationally. Why didn't we care about all the other endangered birds and wildlife that we were supposed to be sharing our planet with? I became particularly concerned about the real-life plight of the Eskimo Curlew, an enigmatic wader that is now almost certainly extinct, with the last validated sighting as long ago as 1963 and a possible sighting of a flock of 23 in Texas in 1981. If the Ring Ouzel had become the bird that I dreamt of seeing in the UK, then the Eskimo Curlew was the bird that I most wanted to see in the world. I believed then, as I still do now, that I will live to see one.

57

Not a happy occasion – starting secondary school.

BATMAN AND ROBINS

I hated starting secondary school. I had spent the summer holiday dreading it. Especially after previously enjoying a relatively cosy existence, being able to walk around the block to get to primary school and then come home to my back garden nature reserve. Life then was perfect. Getting ready for my first day was not an enjoyable affair as the earlier sunrise over the gardens, distant rooftops and factories was far more interesting.

During the summer break I had invented a TV programme in my head called 'Flyways' featuring me as both cameraman and presenter, giving my imaginary viewers a live commentary on the movement of birds over my house. I even invented my own theme tune that was loosely based on the popular children's TV series *Magpie*, with the remaining incidental music being provided by the radio. Watching gulls flying over against a blood-red sunrise with David Bowie's *Star Man* playing in the background was priceless.

In my daily show I had noted that there were more Feral

Pigeons heading towards central London during June and July than at any other time of the year. I put that down to migration. I used to report on the continuous streams of Starlings heading to central London during the evenings. I had worked out that thousands were journeying to their Trafalgar Square roost. Ironically, just a few years later their population was to fall through the floor and the spectacle of watching the flocks heading into town became consigned to memory.

Migration was a concept that I just couldn't understand. For many years I believed that the autumn migration period lasted for just one month and one month only – September. It was the time that I upped my backyard surveillance to record migrant Chiffchaffs, Goldcrests and Swallows, together with a host of unidentified flying objects, plus a few unnamed creatures flitting around in the bushes. Once October rolled in I assumed that it was all over. Well, there was no more leisurely annual September vigilance once I was dropped off outside the gates of Cardinal Hinsley Catholic Boys School on 2nd September 1974. I had to leave home earlier to get to wretched secondary school. Cardinal Hinsley was situated in Harlesden, a relatively alien territory for me, being around two miles south-east of my Wembley home. I was raised as a Catholic, but even so, the prospect of attending a school named after a leading Catholic dignitary who I had never even heard of was daunting. It sounded heavy duty and the prospect of being taught by priests and nuns filled me with dread. Surely I would end up being damned?

The first week was hell on earth. I had no friends apart from Frankie Thomas, an Oakington Manor schoolmate and a rather wild boy who lived on my road with his alcoholic mother. He was a troubled lad who loved to cause havoc, and although we got on well I didn't really fancy hanging out with him. So at break times I would go to the end of the playground and watch the passing traffic outside the school. I would stand there close to tears, watching out in the vain hope that my dad would come past to pick me up and take me home. I wasn't sure about my classmates either. It was also weird being in a school with no girls.

61

I needed an ally. By the second week things had improved and I was beginning to forge friendships. One boy who attracted my attention was my classmate Alan McMahon, born to Irish parents who had arrived on British shores at around the same time as my parents. Alan interested me because he lived literally five minutes' walk from my home in Monks Park, almost next door to the library from whence my stolen field guide had originally come. His biggest selling-point, apart from the fact he had a brown belt in karate, was that he had a self-professed interest in science. He was the chosen one, so one day I challenged him by stating that if he was interested in science then he had to be interested in birds too. Of course, he had no interest in birds but he was nevertheless curious, and to his eternal credit he agreed to me giving him a crash course in birdwatching.

That weekend I called round his house and met his mum, dad, brother and sister for the first time. I was not aware that he was about to have tea with his family, and just thought that we were going straight out and that he had gone to fetch his trainers. My understanding of how to behave in someone else's house was rudimentary to say the least. On seeing the fine display of food that Alan's mum had put out whilst waiting in his front room, I calmly pulled up a chair at the table. His horrified mum could only stare aghast as I brazenly asked if I could have a scone with my cup of tea that had not even been offered yet. Before she could answer I had already helped myself to a delicious-looking scone and was slapping strawberry jam on it. Talk about starting off on the wrong foot!

After that dreadful faux pas Alan and I headed out for his inaugural birding session into Monks Park itself. Although the parish was called Monks Park I always associated its name with the municipal park that was just beyond his garden. The area was a thin strip of parkland of about 16 acres, stretching from the junction of the busy Harrow and North Circular Roads in the south-east to within the shadow of Wembley Stadium to the north-west.

It contained all the usual things you'd expect to find in a park: a cricket pitch, a couple of football pitches and an incredibly dangerous and, as a consequence, very popular playground with huge swings, a slide whose exposed bolts would snag your clothing and a wicked witch's hat. All of which should have carried health warnings as many a child crawled away from that playground with anything from grazed elbows or knees and twisted ankles to lost teeth and even broken bones. The injuries were usually received through jumping off the swings from a great height, or from falling off the witch's hat while overcome by self-imposed dizziness after being spun around too quickly by your mates. It was a rite of passage. You expected to get hurt. Those were fun days.

63

The rest of the park, beyond the playground and towards Wembley Stadium, was far less visited and thus turned out to be the best area for birding. There were a few older trees, although nothing that constituted a wood, and a bowling green surrounded by a thick hedge. The concrete-bottomed River Brent bordered the whole of the eastern edge of the park and could only be reached by scaling the concrete fence posts that were lined with thick brush. Acres of derelict ground lay on the other side of the river, which either raged like a wild Amazonian torrent or trickled along its fetid concrete course, which was embedded with abandoned shopping trolleys and assorted car parts, as it did during the famous drought of summer 1976. But it was the wasteland area that was to become our happy hunting ground.

On our first visit I happily showed Alan his first Mallards as they swam uncomfortably around the garbage-strewn river. I watched his face as he craned to see Swifts wafting overhead and I felt proud of myself when I confidently introduced him to the Greenfinches that I had identified by call alone as they commuted over our heads. But it was the male Bullfinch I found which really resonated with him. He just could not believe that such a brightly-coloured bird could live in such close proximity to his home. I smiled smugly when he said that he had genuinely found our first session interesting but I needed to get him hooked.

So began a period of continual badgering and canvassing, both in and out of school, designed to wear Alan down until he became a birder. I would call for him most weekends, and sometimes during the week, to talk birds and visit our patch. It was not until the following spring that I realised he had gone through the stages that started with being mildly interested, graduating through to being keener once he was out in the field. I would have been happy with that but he didn't stop there. Like a moulting process, he went a stage further by getting a field guide and a pair of binoculars, and then he even started popping out to bird Monks Park on his own. My selfish plan had succeeded; I had finally got myself a birding buddy, albeit through coercion.

My notebook collection was building up. Most were filled with list upon list of things that took my fancy. Aside from the endangered birds and my world list, I also started a list of mammals of the world, plus country-by-country lists based on the distribution maps in my stolen Heinzel, Fitter and Parslow guide. I spent so much time writing that by the time I was 12 my joined-up writing was pretty neat and awesome. But it was my note-taking that really began to develop in those early years. I wrote about the birds that I noticed in my garden, describing how the Woodpigeons would descend into my parents' vegetable patch to voraciously peck at all the runner beans, and reporting on the Blackbirds that nested in the hedge separating our garden from next door's. I noted that when I gently shook some of the twigs near the nest, the youngsters would raise their scraggly wobbly heads skywards with beaks wide open, making their shrill begging cries.

Over at Monks Park, I had begun to make daily lists of the species that Alan and I were finding. These lists had the birds we discovered written out in the same family order as my stolen field guide, a systematic order that I have broadly stuck with to this day, despite the subsequent changes in thinking regarding certain families and their relationship with other groups of birds previously thought to be closely related. I mean, I cannot accept the current school of thought that implies that ducks, geese and swans are at the beginning of the list. For me it always will be the primitive-looking divers, or loons as they are called in North America.

The next couple of secondary school summer holidays were spent hanging out with Alan in Monks Park on a near-daily basis. We used to hide our binoculars when in the park for fear of victimisation by some of the more unruly kids that frequented the area. Most of them usually stuck around the playground in the centre of the park. Once we reached the north-western area things were invariably quieter, with our only disturbance being dog-walkers or some guys playing cricket. We made a series of camps in the scrubby waste ground across the river, where we imagined ourselves as explorers of new and undiscovered frontiers. We also had a small camp on the park side of the river in some bushes that lined the concrete fence. Our little hideaway was behind a bench that was set almost flush against the fence. We used to sit there totally invisible to anyone in the park, with our legs dangling over the edge of the fairly high concrete riverbank. We would chat about school, discuss birds and sometimes we would jump down to the river to try to catch sticklebacks.

One day, as we were sitting in our riverside camp, a dog walker plonked himself on the bench directly in front of us. He was so close that had he broken wind we would have been in trouble. He was totally oblivious to us as we stared at each other, trying not to giggle out loud. Suddenly, his golden retriever showed up and cottoned on to us immediately. He started frantically digging and viciously growling as he tried to get to us. We froze. His owner repeatedly shouted at him to stop before eventually getting up and dragging away his peeved pet.

We were largely on our own when we visited the waste ground. However, on a couple of occasions we had to beat a stealthy retreat when we came into contact with the gypsies that occasionally passed through. I'm sure that we had nothing to fear from the people that we saw, but we were always aware that we were away from the relative safety of Monks Park, and we felt that birdwatching boys were everybody's quarry.

We read somewhere that we could cook food using a campfire. So we bought a whole chicken and a tin of beans and made a small fire using natural materials, which we started by rubbing sticks together – stuff that would have made Bear Grylls proud. Once the fire had died down, we wrapped the chicken in tin foil and buried it in the ashes whilst cooking beans in a pot over a smaller fire that we started. We would then wander off for an hour to see what birds we could find. On our return we would dig the chicken up perfectly cooked and enjoy our meal. We certainly were proper explorers.

Visiting our camps also gave Alan an outlet for his scientific leanings. We had learnt during chemistry that certain chemicals were combustible. Armed with that information, we started a bigger fire, this time with petrol as the main fuel for the flames. We then planted several aerosol cans into the flames and ran off, taking great delight in the series of explosions that began to ring out behind us. During those summer months we used to watch Skylarks rising high into the Wembley skies. We also discovered a pair of Kingfishers that had taken up residence along the stretch of river near one of our camps. Both were new birds for us, but the novelty soon wore off as we quickly took their presence for granted.

That was the thing that characterised our lives then. We went through a short period of taking things for granted. We thought that we would be kids forever, and that Monks Park and its environs would remain as they were forever. We didn't really understand that there were conservation issues to contend with. We were just living in the moment.

68

During the winter we discovered that we had a regular flock of Tree Sparrows residing in the riparian bushes near the playground. I recognised them after reading an article in *Bird Life* and I remembered seeing the same sparrows with black blobs on their cheeks on our patch. We would count upwards of 70 birds, and after a while we barely gave them a second glance. We were so blasé. Little did we know that a couple of years later our Tree Sparrows would disappear completely, when the land that they wintered on became a housing estate.

I became aware of my juvenile birding ignorance in the spring of 1975, when Alan and I were doing our usual circuit of the park, a circuit that invariably included counting the Feral Pigeons and hunting for sticklebacks in the river. On this particular day I was pointing out a few birds for Alan, flaunting my much-vaunted bird knowledge. Suddenly a small bird flew up from the ground to land on a park bench. I didn't even bother to look at it; such was my confidence that I had already dismissed it in my mind as a sparrow. Alan saw it and asked me to identify it. Before he could finish his sentence I cut him off by calling it a sparrow. He looked at it again and dared to ask if I was sure.

"Of course it's a sparrow", I snorted. He's only been birding five minutes and already he's questioning me, I thought. But still he persisted.

"It doesn't look like a sparrow, Dave. Take a look at it". We didn't swear in those days so I snorted again "Bah!" and put my bins to my eyes. What I saw made me gulp. An adult male Northern Wheatear in full glorious Technicolor. My first. Our first. I was embarrassed. I apologised to Alan immediately. I had learnt a major lesson. From that day on I kept my ego in check and tried to be humble. I looked at every bird twice, even if I thought I knew what it was, either because it could be something different, or doing something I had never seen that species do before. I had entered the age of caution and humility.

Back at school I was re-reading my Heinzel, Fitter and Parslow guide for the hundredth time, this time drawing up a hit list of birds that I needed to see. High up on the list was the Mediterranean Gull, purely because it looked so graceful compared with the Black-headed Gulls I was already used to. I think it was the Med Gull's Persil whiteness and obvious scarcity that captivated me. When was I ever going to see one? Was I ever going to see my ultimate favourite bird, the Ring Ouzel? Would I have to make the long trek to the Yorkshire Dales sometime in the distant future? I'm not sure what originally attracted me to this bird. As I mentioned before, it could have been the sense of familiarity I felt I had with it due to its similarity to the super-abundant Blackbirds. Seemingly the only thing that separated it from its more common cousin was a white gorget and an untameable wildness.

Meanwhile, my meticulous study was beginning to pay dividends, as I was recognising new birds purely by the diagnostic features that I had memorised. During a French lesson I found myself gazing out of the window at a small party of gulls that had settled on the concrete playground. I rarely got distracted during French because our teacher, Miss Nestor, was a young hottie who just about everyone fancied. One of the gulls stood out from the crowd of Black-headeds. It looked like a small Herring Gull, but had greenish legs and a slim yellowish bill with no red spot on the lower mandible. It was a Common Gull – my first. I nudged Alan to show him a new lifer. A burst of excitement welled up inside me as I recalled reading that they only bred in Scotland and were far from common. "Miss!!" I blurted. "There's a rare bird on the playground, can I go out and see it?" Laughter erupted in the class. Miss Nestor looked at me incredulously.

"What bird is it, David?" she enquired. I noted that she didn't even glance in the playground's general direction.

"Err, it's a Common Gull miss…"

"A Common Gull," she said, sarcastically emphasising the word 'common'. "You're staying here. Now get on with your work." Thus ended my first attempted twitch amid the backdrop of raucous laughter.

I made frequent trips to the library searching the paltry natural history section to see if any new books had come into circulation. I picked up John Gooders's *Where to Watch Birds* and found I could not put it down again, as for the first time I was exposed to a list of sites to go birding and what birds to expect. I borrowed the book, never to return it. Reading about the huge variety of what seemed like exotic birds, such as Cuckoos and Little Ringed Plovers, to be seen at a host of sites outside London was too much to resist.

Alan and I were just kids, so we had no means of getting about, but when I showed him the book we immediately started to plan trips to the birdwatcher's Mecca of Norfolk, and beyond. We figured that we could go to East Anglia once we had hit 18, and catch trains to places like Tring Reservoirs in Hertfordshire. As much as I delighted in studying the plumage detail of the House Sparrows in my back garden I yearned to travel further afield. The London section only detailed a few sites but one in particular seemed like the next-best thing to going to Norfolk: Rainham Marshes. Although classed as London the site in fact straddled the border into Essex. The one line from the description that really got our juices flowing was "At passage periods gatherings of waders are often large and exciting." We had little experience of waders apart from the occasional migrant Common Sandpipers we found along the River Brent, and the book's cover, depicting a typical shoreline vista of gangs of Knot, Redshank, Oystercatcher and Curlew, added fuel to our desire. However, it was certainly unknown land and there was no way our parents would allow us to travel so far on our own.

72

I hatched a plan. After secretly working out bus routes and timings within one of our camps in Monks Park we decided that we would travel to Rainham without telling our parents. So one Sunday morning I told my mum that I was going around to Alan's to do some birdwatching in Monks Park. Instead, I cycled around to his house where I stowed my bike and together with our bus map we travelled the whole width of the capital, passing through deepest east London, an area that frightened me somewhat because I felt it had the reputation for being a hotbed of racism. I had heard plenty of stories in the school playground about this being the area that spawned the National Front and skinheads. In my mind it was a very different landscape then, pre the 'Asian invasion', it really was the land of the Cockney rabble.

Nearly three hours, three buses and some 20 miles after

leaving Wembley we arrived at the site and had an amazing time exploring this very urban wilderness. It consisted of dumped motorbikes, cars, abandoned tyres, shopping trolleys and worryingly, spent cartridges at regular intervals within an area of grazing marshes, silt lagoons and a massive landfill site, right next to the Thames Estuary at Purfleet. Later, I discovered that the whole site had been closed to the public for well over a hundred years as it was largely used by the military as a firing range. We were effectively trespassing. Regardless, our inaugural trip was totally mind-blowing with a hatful of new species on our lists including our first Little Ringed Plovers. Best still, we arrived home with our respective parents none the wiser.

73

Our early visits were not without incident. In those days, when visiting this pre-RSPB wilderness, you had to make a choice: enjoy some potentially great birds while warily looking over your shoulder or turn around and go back home. Rainham was a lawless, derelict land frequented by unruly, air rifle-toting Essex boys. In 1976, after catching sight of our first-ever Cuckoo, Alan and I had to beat a hasty retreat after coming under fire from a bunch of hostile local lads. A pellet whistled past our heads as we were literally chased out of town. We visited Rainham a few more times after that harrowing occasion, and on one visit in May 1978 we recorded a range of birds including a singing Cuckoo, Tree Sparrows and our first-ever Firecrest. I thought nothing of this discovery until nearly 30 years later when I happened to read the Rainham checklist and noted that there was no record of this species. We had apparently seen the first one ever for the site.

We also saw a Corn Bunting that day, a species that by the early '80s we were calling 'Essex Birds' because of their apparent prevalence in the county. Our trips to the Essex countryside were characterised by the sight and jingly sounds of these drab-looking seedeaters. Since those heady days their fortunes in Britain have completely changed and their phenomenal decrease has been well-documented.

I'm pleased to report that Rainham is a very different proposition now, with the RSPB's Rainham Marshes Reserve now firmly positioned as the premier birding site in the London area. It boasts an impressive list of species, including some totally unexpected rarities such as a White-tailed Plover from Central Asia, and more recently a Slaty-backed Gull from East Asia – the latter being a bird that only a gull-loving freak could really get excited about.

My mum never suspected that I was journeying across London as a boy to go birding and she was shocked when the truth came out a couple years ago when I wrote a piece for *Bird Watching* magazine recounting the story. However, after the Rainham gun incident Alan and I took the step of 'arming' ourselves against any future attacks whilst out birding. A visit to the outdoor shop on Wembley High Street ensued, and soon we walked out as proud owners of a small penknife and a catapult each. I guess our purchases gave us a sense of security, but obviously it is not something that I would do today if I were a kid. The catapult was Alan's idea. It doubled up as a offensive weapon and a bird flusher. Thankfully, we never needed to use our weapons and they remained tucked discreetly in our birding jackets, apart from once, ten years later, when we used our knives to cut a drowning Long-tailed Duck free from a fishing net at Walberswick in Suffolk.

John Gooders's book opened the door of opportunity, gifting us details and directions to birding sites we had not previously known about. We caught the train and visited Tring Reservoirs, where we proudly saw our first-ever Greylag Geese. We thought we were witnessing truly wild birds that had flown in from the high arctic, not knowing that they were probably hatched yards from where we saw them. We also discovered Staines Reservoirs, on the western outskirts of London. The ornithological history here was legendary and the site had a bird list to rival anywhere in the country, with beauties such as Buff-breasted Sandpiper, Collared Pratincole and Wilson's Phalarope having all turned up, whilst scarce terns and gulls were regular. In addition a host of waders always seemed to be attracted in whenever one of the basins was drained for maintenance.

Staines Reservoirs was also the stomping ground for many of the birding world's great and good. There are in fact two reservoirs (the north and south basins) that lie just south-west of Heathrow Airport, so when birding we were usually under the shadow of a noisily-landing Jumbo. A central causeway with general public access separates the water bodies and the site is now protected due to the important numbers of wintering diving ducks. In the days before bird information services, when the grapevine was king, the causeway was a great place to meet up with fellow birders to chew the fat and to find out what rarities were around. In the winter you were exposed to arctic winds with no cover and in summer your own personal cloud of midges would envelop you. I used to wish that I had an attendant gang of Spotted Flycatchers to take care of those pesky flies.

76

Birding here was initially very hard for us, because we were still beginners and not used to seeing and recognising small dots bobbing around at the far end of one of the basins. The older birders here were generally unfriendly and one group, whom we dubbed the 'Berkshire Bastards', used to go as far as turning their backs on us when we asked them questions. We didn't care, though, as we found their ignorance amusing. Many years later I bumped into one of the BBs at the British Birdwatching Fair. As he chatted to me in front of his mates, I took great delight in reminding him of his previous attitude, and how he was the founding member of the Berkshire Bastards. Watching his face turn crimson was a picture!

CORNELIUS RAVENWING III AND THELONIOUS MONK

In the winter of 1975 I joined a local youth club. St Michael's Youth Club was based up the road from me, appropriately enough on St Michael's Avenue, and was where the local kids used to meet on a Thursday night to play records, dance, chat, play games and eat sweets. As I lived in a fairly respectable area the members of the youth club were by and large well behaved.

Our leader was Mike, a calm and fair man in his thirties who commanded respect from all the kids that used the club. Shortly after joining I became Tuck Shop manager, but I was rubbish at it. I kept on eating all the sweets that I had painstakingly bought from the previous week's takings. The problem was that I was literally eating into the profits, so to augment my stock I used to pinch Mars Bars and the like from good old Woollies. No wonder they went down in the end.

During the summer of '76 I went youth hostelling with a group of kids from the club. Mike had planned a trip to Scotland during the previous winter, and the idea was that after gaining our parents' approval we would all pay a certain amount each month to cover the train fare and accommodation plus some money for food and spending. I had to do a major selling job to convince my dad to let me go, as for some unknown reason he was totally against it. I had to draw upon my trump card and wheel in Alan's parents to talk him around. A visit to their house, a bit of Irish charm and a few Guinnesses soon changed his mind.

78

Finally the day in May came and I was on board the sleeper train leaving Kings Cross, heading for Scotland with a bunch of similarly excited boys and girls. Alan and I shared the same sleeper car in which he sensibly slept. But I was so excited that I stayed awake the whole night, hanging out of the window of the door outside our bunk, birding in the dark. I didn't want to miss a thing. It was my first time outside London apart from a couple of visits to the seaside with my parents, the trips to Whipsnade Zoo with primary school and my trip to Radlett Gravel Pits.

When the train pulled in at Carlisle it was still pitch black. The station guards were loading a carriage with mail so we were stationary for quite some time. I remember hearing a strange incessant churring, reeling sound emanating from the dark nearby and wondering whether in was an animal or some

A 'wee pregnant canary' getting to grips with the birds of Scotland.

odd mechanical noise. Of course, years later I was to realise that I had been listening to my first-ever Nightjar.

We eventually arrived in Inverness and made the journey south to Aviemore, where we were based. We spent the next week walking the expanses of the Cairngorms National Park. I was strolling around in a bright yellow cagoule and a

79

miniature Michael Jackson afro making me look like a stylistic disaster zone. No wonder the locals stared at me. With my oversized binoculars under my raincoat I looked like a 'wee pregnant Canary', as a local in Aviemore accurately described me. I was stared at quite a lot whilst in Scotland, but then again, I suppose I was the only black person for miles around. I never saw another black face the whole time I was there, which at first was a little perturbing but I soon realised that the attention I attracted was out of curiosity, not racism.

However, those stares soon changed to looks of disbelief when I produced my binoculars. But I didn't care because I was in heaven, seeing birds that I had never seen before. Alan and I were the only birders in the group and were overjoyed when we heard flocks of Wigeons whistling in the mist and when I flushed a proper Nightjar from the woodland floor in the Craigellachie Forest. It was a complete surprise to me and I had no idea what the strange brown bird I disturbed was until later on that evening when I consulted my field guide and remembered that it landed along the branch of a tree as Nightjars classically do, as opposed to across it like most other birds, before promptly blending into the background.

Buzzards were all over the place and they were totally new for us too. We watched them for hours, trying to work out why they were not Golden Eagles – another bird we hadn't seen before. Buzzards looked massive to our untrained eyes as they flew around the glens, frequently catching us out. One afternoon after a long walk through some nameless part of the Cairngorms I saw a huge raptor soaring high overhead. I got Alan onto it straight away and we instinctively knew that we were looking at our first Golden Eagle. On reflection, the huge rectangular wings were a giveaway.

However, due to our newly-installed caution we went against what we thought and asked Mike for his opinion. I guess we were harking back to the notion that all adults, especially tall ones, knew more than us about everything. Mike was really tall. He dismissed our claim, saying it just wasn't an eagle. Unbelievably, we took his word, the word of a non-birding grown-up. It wasn't until I saw another Golden Eagle in the '90s that I realised that the original bird was one, so I awarded myself a nice back-dated armchair tick.

81

The following year, Alan and I went back to Scotland youth hostelling with Mike and some other youth club kids. This time we journeyed further north in a sweltering heatwave, along the beautiful west coast from Ullapool to Cape Wrath. Our attitudes had also changed. Punk had just broken onto the music scene and we were enveloped in it. Our favourite bands were The Sex Pistols, XTC and The Stranglers, so we spent loads of time chilling in the Scottish sun listening to our music on Alan's cheap plastic tape player that he had won at a recent school disco raffle. He was torn between the first prize of a Showaddywaddy album or a tape recorder. Fortunately, I persuaded him to take the second prize.

In complete contrast, I had developed a great love for the Electric Light Orchestra, after recently hearing their music being played on Kenny Everett's show on Capital Radio. I was a big Kenny Everett fan. So with such wide musical choice, plus a tape of various bird calls, we were in our element. Our new attitude was steeped in typical teenage angst and was reflected in our new birding attire. Gone was my canary-coloured raincoat; in came the black trenchcoat, 16-hole Doctor Martens and dark trousers and t-shirts. We also invented pseudonyms for ourselves; Alan was Cornelius Ravenwing III, and I became Thelonious Monk. The Ravenwing name was completely made up, whereas I had vaguely heard of Thelonious Monk, but because Alan hadn't I ran with it. For the rest of our schooldays and beyond we would go on to leave notes of sightings in many a bird hide diary under our assumed names. It was a while afterwards before I realised that I had indeed taken the name of a spectacularly talented jazz musician!

We were much more ornithologically experienced during this trip, picking up on species that we had missed the previous year. We quickly added a host of new birds to our tally including flocks of gorgeous Twite, Dunlins sporting their black summer bellies, elegant Arctic Skuas, some 'real' seagulls – Kittiwakes – and perhaps surprisingly, our first ever Cormorants. Back in the mid-1970s Cormorants were not the urban bird they are today. We had to wait until the 1980s before we started seeing them regularly in London.

The most heavenly place we visited was Achmelvich, a wonderfully peaceful hamlet on the west coast, not far from the Summer Isles. We would step out of the hostel in the early morning to the sound of silence, apart from distantly-calling Cuckoos. A short walk over a tussock-covered sand dune and we were suddenly on a glorious, deserted, golden beach with silent waves lapping onto the shores that stretched for miles. It was idyllic. I was surprised that such a place existed in Britain, as I had always thought that Scotland was a cold and wet place. We were eight miles from the nearest shop, so we were truly isolated. Alan and I spent what seemed like years wandering around the countryside chatting, watching Swallows swooping around and seeing soaring Buzzards. Once, we happened upon a serene-looking loch and started scanning it from the roadside. We were soon ogling at yet another new species, some summer-plumaged Slavonian Grebes with their resplendent orange ear tufts really kicking out in the bright sunshine. The only problem was that there were some pretty beefy-looking Highland Cattle with mega horns standing in the way between the grebes and us. To be honest, we were scared of the cows but wanted to get a closer view of the grebes.

After deliberating for nearly an hour we decided to venture into the field, as any good pioneering naturalist would. It was hair-raising as all the cattle stopped chewing and were staring at us intensely. It would have taken literally five minutes had

84

A bit of birding in the true wilderness made a refreshing change from Monks Park.

we just walked straight across but instead it took us another hour to delicately circumnavigate the beasts. After that detour we got to the water's edge, only to discover that the grebes had disappeared. Gloomily, we looked at each other and then at the shaggy long-horned monsters behind us before inching our way back to safety.

A few days later we went on a trip to Handa Island Nature Reserve, farther to the north, where we experienced seabird colonies for the first time. Puffins, Guillemots and Razorbills abounded and we registered our first-ever Arctic Terns which were nesting everywhere. Later, we came to a bay where we noticed a group of Eiders variously lounging on the rocky shoreline and bobbing up and down on the surf. They were a new species for us so we stopped to watch them.

Eventually, we noticed a strange dark duck in the bay with whitish markings on its face. We just could not work out the identity of the mystery duck so we made copious notes and pinned it on the warden's locked cabin door. Several weeks later we received a letter from the Handa Island warden congratulating us on discovering a Surf Scoter, a really rare bird. We had found our first national rarity, and despite seeing a Surf Scoter years before seeing a regular Common Scoter, we were baptised. We had arrived.

The drought of the summer of 1976 was a memorable time. Hosepipe bans were rife, the sun was baking hot, and during the summer holidays I lived in a pair of shorts. It was also the summer that I conducted my third and final daring dawn raid down at the local library. I was looking through the still meagre natural history section for new books when I came across *Birds of Town and Suburb* by Eric Simms. Essentially, it was a book about urban birding in which, according to the sleeve notes: '...Mr Simms describes and explains all the bird life of British Towns, starting with the inner rings of towns and working outwards'. Little did I know at the time but this book would become my urban birding manual, teaching me that birds and other wildlife often deliberately seek out habitats in cities, and that as a consequence they should be expected and, indeed, looked for. I discovered that birds actively flew over cities and not around them; attracted to the green areas such as parks and gardens, following watercourses or perhaps even retracing ancient migration routes.

Eric Simms talked of his experiences within different urban environments, discussing the behaviour and distribution of the various species he encountered. Although he spoke about Britain as a whole, mentioning towns and cities I was yet to experience, he also wrote about urban birding in London. He made me realise that I was right to think that anything can turn up anywhere when he mentioned watching a Whimbrel flying over his west London home in Ladbroke Grove in the '50s, and the Red-breasted Flycatcher he discovered foraging in the trees of the communal gardens in Ladbroke Square. These were all real events that could happen to me. That thought filled me with inspiration and unconsciously made me even more observant, ever on the lookout for the unusual amongst the usual.

But there was another major thing that he introduced to me that changed the course of my birding life: the concept of local patch birding. Many of his observations were centred on

86

the areas he had constantly watched over long periods of time. He spoke of seasonal changes in populations and species, making me consider for the first time that migration was a very fluid event, not just confined to September. Waders started moving from mid-June onwards, migrating down from their high arctic breeding grounds to fly over our cities en route to their wintering areas. Some of our summer migrants such as Cuckoos and Swifts started moving out from as early as July while the bulk of the summer visitors were on the move from March to June, and from August through to November, crossing over with our winter visitors. This was a notion that I had not thought about before. He also kept referring to the birds of the allotments in Dollis Hill and at the nearby Brent Reservoir. I reached for a map and realised that the reservoir was very close to me. Soon I had convinced Alan that we needed to broaden our horizons and visit this new site.

87

Brent Reservoir and environs have been in existence since it was first constructed in the 1830s. It was originally known as Kingsbury Reservoir and colloquially as the Welsh Harp, a name I originally thought was due to its shape but which I have since learned came from a local pub, the Old Welsh Harp Tavern, that stood nearby until the early '70s. Of course, in those early days it was situated in an expanse of countryside and was used as a hunting ground and for recreational purposes. A couple of those early hunters were James Edmund Harting and Jock Walpole Bond, two ornithologists who recognised the site's value as an important refuge for birds, and duly collected some pretty interesting specimens.

The 19th century ornithologist Harting was probably the best authority on British birds of his day, while Walpole Bond, who lived from 1878 to 1958, had the dubious honour of being one of the top egg collectors in the country. He claimed to have seen the eggs of every single regular British breeding bird, in situ. Collectors such as these procured rare vagrants at Brent Reservoir, such as Little Bittern, Squacco Heron and White-rumped Sandpiper. Perhaps the most contentious species to be 'obtained' was a pair of Purple Martins, discovered during September 1842. This North American swallow had never appeared in Britain before, but the records were never accepted and the species remained a pipe dream for British twitchers until an immature was found at Ness on the Isle of Lewis, Outer Hebrides, in September 2004. I do like happy endings.

Over the years the area became eroded by the expansion of the suburbs to become a truly urban site encased by housing, industry and busy arterial roads. Despite this, Alan and I were in heaven. At last we had found a real local patch that had reedbeds, water, woods, marshes, fields and expanses of sky, and unlike Rainham Marshes was just two bus rides and 40 minutes from our homes.

We began splitting our time between Monks Park during the week after school and Brent Reservoir at the weekends. I voraciously read up on the bird history of the site and realised that we were onto something great. Our first few visits resulted in a hatful of ticks including a lot of the basic species that any beginner would see: common birds such as Great Crested Grebe, Shoveler, Fieldfare, Redwing and Reed Warbler. Skylarks bred in good numbers and we even had a colony of Tree Sparrows that bred in the nursery at the eastern end by Birchen Grove. We used to get there at dawn, a habit we got into from an early stage having realised that it was the best time for birding and to avoid being verbally molested by the rogue boys that seemed so widespread in those days.

One winter morning we bumped into a couple of older birders who were looking for Smew, a Scandinavian duck that at that time chose Brent Reservoir as one of its main wintering areas in London. Gatherings like that are now sadly a thing of the past.

The Smew's heyday was back in the 1940s, when at least 40 birds were to be found. Knowing how scarce they are in the UK now it was mind-boggling to think that 144 chose to reside at the reservoir during the winter of 1956. In the mid-1970s you could have pitched up and found as many as ten of them on the water, and people used to travel here by the coachload to witness these beauties. We took our Smews for granted and calmly told the guys where to find them. As an afterthought, one of them told us that they had just been watching a Snipe feeding out in the open. We had never seen a Snipe before and excitedly asked for directions. I knew exactly where they had seen the bird and as Alan turned to verify their exact location with me I was already 100 yards away, sprinting towards a certain tick.

Brent Reservoir was where we really cut our teeth as birders. We made many mistakes and misidentified countless birds. Other birds were consigned to the 'ones that got away' bin such as the certain Marsh Harrier that we saw in September 1978 that I described in my notebook: 'a Carrion Crow mobbed and chased away a very large bird of prey about the size of a Buzzard but its wings were longer and narrower. It was brown in colour.' Earlier that year we happened across a Barnacle Goose swimming amongst the Mallards. I wrote in my notes 'At first we thought it was a Canada Goose (a scarce bird in those days) but upon closer inspection it turned out to be a Barnacle. At first we weren't sure if it was wild but after a few tests e.g. throwing bricks at it, we found that it didn't have any rings on its feet.'

We were both very interested in the science behind birds but went about exploring it in a very crude way. In the summer, during school sports practice at the park, we noticed that Swifts would often swoop down low, no doubt feeding on the insects that we were all disturbing as we ran after rounders balls. Our plan was simple. We slipped our trainers off and in the last minute would chuck them in the air in the hope of knocking a Swift out of the sky so that we could examine it before releasing it again. You will be pleased to learn that we were miserably unsuccessful, the Swifts easily out-manoeuvring our bulky Adidas footwear.

At school, Alan and I were beginning to move in different circles, but birding was still our major bond. But we were finding more and more reasons not to go birding, ranging from being too lazy to get up in the morning to choosing to spend the time doing other things. Discovering drink, girls, music and a few more girls became the order of the day. *Saturday Night Fever* had just come out, and most people were going disco-crazy. I started going under-aged clubbing, learning about girls either in the clubs or down at the local convent. As Alan concentrated on his academic studies I majored on deejaying, social intercourse and football. I became more distracted and birding started to take more of a backseat in my life. Attitudes towards birds and nature from other people reached their lowest ebb at that time, and I buried my passion deeper than ever. It was seen as being wimpy and you were automatically treated as a lesser being if you displayed any leanings towards nature. Although my radar was never switched off, and I was continually noticing birds, my birdwatching would often be restricted to seeing something from the corner of my eye while in conversation and leaving it at that. As a consequence the public birdbrain persona subsided and a lot of my friends either forgot or didn't realise that I had an interest in birds in the first place.

91

By 1979 Alan and I had officially been birding just twice that whole year.

My desire to go out birding took a further battering when I arrived at Brent Reservoir by myself one Saturday afternoon with my oversized binoculars in a plastic bag, the same pair I'd had since I was eight. It was a sunny afternoon, so there were a few more people around than usual and quite a few yachtsmen were sailing on the reservoir. I was dimly aware that a man was following me but I assumed that he would walk past me at some point. He didn't, and had soon caught up with me. He asked me about the contents of my plastic bag in a very effeminate way. I felt uneasy but told him that I had a pair of binoculars and that I was birdwatching. He then continued talking to me as we headed towards a rather dense Weeping Willow. He lured me under the tree's canopy where he proceeded to masturbate in front of me. I was confused and I suppose a pretty naive 14-year-old because for a couple of minutes I thought that he was trying to have a pee. It was only when he asked me if I wanted 'some fun' that common sense suddenly kicked in and after swearing at him I bolted and kept running with my heart in my mouth. I didn't speak about that incident for several years as it had genuinely scared me. It was not until I was around 18, and a bit better-versed with the ways of the world, that I was able to process the situation and move on.

Alan and I stayed on at school until we were sixteen in the lower sixth year. Our heady summer days watching birds in

Monks Park were never to return as we were seriously drifting apart. When we left school Alan went on to higher education and I got a job at Neasden Tax Office as a clerical assistant. My birding instinct was still heavily sedated as I involved myself in learning about people in the world of work. I guess there was the distinct possibility that my interest in natural history could have died away completely, evaporating as quickly as it came. But it was innate, a basic part of my makeup. I was experiencing a typical adolescent phase, which on reflection was a very important one to have gone through. Had I not got involved in other interests, and immersed myself in other lifestyles, I would probably have grown up to be a square anorak harbouring a secret collection of bird books in my bedroom.

93

In 1981 I went on an organised two-week soul music holiday to Cap d'Agde in the south of France with some of my friends. It was a soul, funk and disco event put on in a holiday complex that was the French equivalent of Butlins and filled with hundreds of British soul music fans, all of whom had made the long journey by coach. In effect it was one of the early forerunners to the '90s Ibiza scene. We spent two weeks dancing to our favourite tunes, chatting up girls and drinking. I virtually became an alcoholic for the period, some mornings guzzling a bottle of rosé in bed before heading out for breakfast.

One afternoon I was stumbling through the site when I came across some Crested Larks on an area of waste ground.

I identified them, despite the fact that I could barely see straight as I struggled to focus on them perched on clods of earth. I remember wishing that I was sober enough to really take them in. I decided then that the next time I went away with my mates I would ignore peer group pressure and stay off the sauce. Those birds have remained the only lifers that I have seen under the influence of drink.

When I came home and sobered up I contacted Alan again. It was the first time we had seen each other in a few months so we sat and chatted about the old times and birds. Before long we decided that we would bust back onto the birding scene with a big bang, aiming to end our apprenticeship and become birding experts. In 1982 Cornelius and Thelonious were back in business, dressed in trench coats and trilbys, and once again many a bird hide wall and diary were scribed with records of Ostriches flying south on migration.

Initially we started off gently by reinstating Brent Reservoir as our local patch, before venturing into the hinterland of Kent and East Anglia and beyond. By this point Alan had passed his driving test and bought a car, ensuring our freedom to go birding wherever we wanted. Alan did all the driving but was not always happy about it as he often got tired quite quickly. A woman got chatting to us one day when we were in a hide at Minsmere in Suffolk, and upon asking who drove, Alan curtly replied 'Yes! This decade it's me and in

94

My long-time birding companion Alan McMahon, aka Cornelius Ravenwing III.

the next decade it will be Dave's turn!' We travelled around almost exclusively birding together until the end of 1989. Alan met a woman and decided to move in with her. It proved to be the beginning of the end of our regular partnership and things were never the same after that. Although we have always remained friends Alan effectively hung up his binoculars, and once again I was birding alone.

SHAPER MAKERS

As you now know, I inexplicably became interested in wildlife, and particularly birds, with my initial interest growing and developing without any guidance. But in life everyone needs guidance, and role models, to help you forge your path. As a kid I remember watching episodes of *This Is Your Life* with Eamonn Andrews, and seeing some nobody stepping onto the stage after being flown all the way from Australia to see their now-famous friend on their special night. The celebrity would humbly speak about the person, saying how much of a mentor they were. Mentors don't have to be peers at the top of their game, who you look up to as inspirational people – they can come from anywhere. I have the greatest respect for people who have humility, are gifted and willing to give people their time and the benefit of their experience. The people who have had an effect on me in my birding seem to share a lot of the same characteristics: they are social beings possessing the common touch; they have intelligence, wit and an element of eccentricity; and they all seem totally down-to-earth, nice people. Those were the

An ever-increasing shelf full of notebooks is part and parcel of being a birder.

qualities that I most admired when I was younger – even if I could not articulate it.

A lot of people would cite David Attenborough as the ultimate natural history role model. I am no different as the man is a giant, a don. But although I watched him whenever he was on when I was growing up, I didn't have aspirations of being a TV presenter. That came much later in life. It is just that David Attenborough is a god – and then there is everyone else.

When I was a youngster, the first naturalist I came into contact with was Gerald Durrell. I have already mentioned that Gerald inspired me to write notes, and gave me a romantic dream of being an explorer, travelling across the world in search of weird and wonderful animals. It's a romanticism that many of us carry in our hearts but very few of us ever actually realize. I guess what I learned from reading his books was that having an interest in nature need not mean that it has to be boring and technical. It was all about fun and adventure. It felt as though anyone could get out there and do what he did. Reading *My Family And Other Animals* I identified with his account of being a child in Corfu, as he was of a similar age to me and had a great inquisitiveness for wildlife. The fact that I could not relate to his middle-class background, his lifestyle on Corfu and the whole slew of pets he had did not matter; it was his love of wildlife that shone through for me. Reading his books stoked the fires of my interest and kept me going until I stumbled across the next major influence in my life, Eric Simms.

For the first time in my life I had encountered someone, albeit through the pages of his books, who was interested in birds to an astounding degree. Eric, who sadly passed away in 2009, was born a Londoner in 1921. He served in the Second World War and was awarded the Distinguished Flying Cross while serving in RAF Bomber Command. It was after the war that he began making his mark on the world of ornithology.

He got a job at the BBC and eventually became a wildlife sound recordist, before going on to make thousands of radio broadcasts and hundreds of television appearances, as well as being a prolific writer of books.

One of the many things I loved and admired about him was that through his writing you could tell that he was truly passionate about bringing wildlife to a wider audience. He wrote brilliantly researched and captivating books such as *Woodland Birds*, *British Thrushes* and *British Larks*, *Pipits and Wagtails,* that to my mind were for nature aficionados. What made them very readable for me was that interspersed with the science were his personal anecdotes. He had a great way of throwing in a simple sentence that almost seemed like an aside, but revealed the immense knowledge this man had acquired over years of watching birds. Speaking about the Blackbird's hunting technique in *British Thrushes*, he said 'I once watched a Blackbird at a range of only two feet from a ward in the Middlesex Hospital in Mortimer Street as it looked for worms in a window box, and it certainly gave the impression of listening as well as looking.' A simple observation, in an ordinary situation, of a common bird that practically anyone in Britain could have seen. But it was enough to make me want to rush to the nearest window to watch and study the first bird that I saw.

He made the ordinary and everyday interesting, especially in one extraordinary work, *The Public Life Of The Street*

99

Pigeon. Although I had heard of the title, I never saw a copy until after his death. It is a fantastic monograph on perhaps the most familiar avian creature in the land, written in a very easy-to-read style that would have enthralled birders and non-birders alike. Had I read it when I was younger it might have changed my attitude towards Feral Pigeons a lot earlier. If only my local library stocked it. Of course, one of my favourite books of all time was *Birds Of Town And Suburb*. It was, as I have already mentioned, my introduction to the world of urban birding, written by the original urban birder. As I grew older I felt as if I was echoing Eric Simms's life.

Well, his addresses at least. He lived in Dollis Hill whilst he was studying the birds of Brent Reservoir. I moved there in the mid-1980s. He resided in Ladbroke Grove. I moved there in the '90s and have remained more or less in the general area ever since. He ended his days in Lincolnshire, a county I have yet to get to know well – but there's still time.

In the early '90s I approached a publisher with the idea of writing a book on the birding sites of London. On getting the commission my co-writer and I sat down to discuss who we would ask to write the preface. I had Eric Simms in my mind from the word go and selfishly canvassed for him to have the honour. To be fair, my co-writer agreed and that summer I had the enormous pleasure of meeting my hero at the British Birdwatching Fair. I felt as though I had just met Prince in Minneapolis. Gushing like a babbling kid, I asked Eric if he would be up for writing our forward. Obligingly, he agreed like the perfect gentleman that he was. When I heard that he had died in 2009 I was really upset, and immediately wrote about him on my blog. I also gave a quote to an online obituary. My words were read by Eric's daughter who contacted me and very kindly gave me his desk copies of *Birds of Town And Suburb* plus his pigeon title, both replete with his actual notes and review clippings. It was an enormous honour. If my house were on fire they would be the two things that I would fetch first.

During the summer of 1984 I had met a young Yorkshire lass out birding with her dad in Norfolk. She was just 14, and a spotty little thing, but was a very keen birder. To be honest, I didn't take much notice of her while I was asking her father about the location of a Barred Warbler that had been seen at Holme. Three years later my birding buddy Alan and I were on the Isles of Scilly sampling the twitching scene for the first time when we bumped into her again, this time with an older man who was clearly mentoring her. Her name was Jo Thomas and little was I to know that she would enter my life again at a later stage and in a major way.

The man accompanying her was Peter J Grant. I recognised him and knew that he was the ex-chair of the British Birds Rarities Committee and author of several books that would eventually include co-authorship of the hugely successful *Collins Bird Guide*. He was an enormously

101

respected figure within the birding community. Jo introduced me to him and we got on immediately. Within minutes he was explaining the differences between Chiffchaffs and Willow Warblers in simple terms that made total sense. I stuck with him and his young protégée as much as I could during the week I was on Scilly, hanging on his every word and asking a million questions. I had absolute respect for him and did not doubt anything he said, even if it flew in the face of popular thinking.

We had picked a bumper year to be on Scilly with many amazing vagrants being recorded, including Britain and Ireland's second-ever Philadelphia Vireo, an Eyebrowed Thrush and a Bonelli's warbler that was latterly assigned as an Eastern Bonelli's Warbler, which is still a species that is recorded extremely rarely in the UK.

102

One afternoon, Alan and I joined a throng busily watching an *Acrocephalus* warbler that was showing strong characteristics of being a rare Marsh Warbler, a bird neither of us had seen before. It seemed to have light, flesh-coloured legs and its plumage appeared to have a colder brown tone than the common Reed Warbler, which are both good indicators. We were congratulating ourselves on our new lifer when Peter arrived. He took a long, considered look at it and announced to the astonished crowd that it was nothing more than an aberrant Reed Warbler. Of course not everyone accepted his prognosis but I immediately started to study the bird more closely, trying to work out why he thought that it was the commoner species. Later when I spoke with him about it he told me to always question the identification of a bird, no matter how straightforward it may seem. Question why it is what people say it is, and don't be afraid to go against the grain. Wise words that have shaped my birding to this day.

I would like to think that I impressed him during our short time together. When we originally met he seemed to raise his eyebrows at me a few times, perhaps surprised by my depth of knowledge and sheer enthusiasm. A Hermit Thrush was discovered on St Agnes, and being based on the main island of St Mary's meant that we all had to charge to the dock – an army of green-clad, optic-wielding men – to load up on the boats to take us the short distance to Agnes. I got myself a seat at the very end of a boat filled with around 30 chatting twitchers and sat patiently with Alan waiting for the boat to begin the journey. As we drifted in the dock area I became aware of a bunch of gulls variously flapping and landing on an area of water less than 50 feet away from us. I checked them through my binoculars and suddenly realized that I was staring at a first-winter Iceland Gull calmly sitting in the middle of the mêlée. Its biscuit-brown upperparts, white wing-tips that tapered beyond the end of its tail and its gentle-looking rounded head were diagnostic. It was also a tick for me.

Realizing that I had found an interesting bird, I broke my gaze to see if anyone else on the boat had noticed my find. Everyone was chatting amongst themselves as before, apart from one man at the opposite end of the boat, Peter Grant. He had seen the gull too and was already looking at me when I saw him. The knowing smile and semi-wink he gave me almost made my heart burst. I felt like I was Luke Skywalker after finally getting the nod to join the Jedi club. I levitated off the boat when we arrived at St Agnes and joined the assembled crowd of several hundred people waiting by some bushes inside a stone-walled field.

The Hermit Thrush had last been seen there briefly a few hours earlier, so it was now a waiting game, with the crowd vociferously recounting twitches of old, cracking jokes and greeting old pals. After a couple of hours staking out the American thrush, I turned to Alan and suggested we go off birding and come back later, as I was completely bored by this point. Our little jaunt resulted in us finding a Melodious Warbler and a Red-backed Shrike.

We returned some hours later to find the same people standing at the same spot. We also found Peter Grant with his pupil in tow. After a brief discussion Peter suggested that we break the mould and go to the opposite end of the field to look at the bushes there. His rationale was that with the amount of people and noise in the area where it was last seen there would be no chance of it staying there. We got to the other end of the field, and within five minutes our quarry jumped up on top of the wall to show itself to us before slipping off again. Peter was right not to follow the crowd.

We all left the Scillies at the same time and I have fond memories of playing birding games with him on the train back to London, such as a round-robin of naming as many birds with white rumps as possible or birds named after people. It was one of the last times I ever saw him, because in 1990 he died suddenly of stomach cancer, at the age of just 47. He was perhaps the greatest ornithologist I had ever met, and I vowed at the time that if I ever got even a tenth of the knowledge he had, then I would devote my life to passing it on to others.

Two other people also had a great affect on me, in a subtler way, but no less life-changing. They added another layer to my passion that I never knew existed. I originally met Rupert Hastings in the '80s when Barn Elms, the site of the current London Wetland Centre, was my local patch. He was such a lovely man and a well-known figure on the London birding scene. A brilliant birder and prolific bird-finder, he was the Surrey Recorder for the London Natural History Society's Ornithological Section, now conveniently renamed the London Bird Club. He was also holder of the capital's biggest bird list.

Rupert was a meticulous and thoughtful birder who never

106

As well as a hatful of rarities, trips to Scilly have also produced a whole host of other special birds, including my first-ever Ring Ouzel.

left any stone unturned. He always spared the time to chat and welcomed me to what was essentially his stomping ground with open arms. I would see him at Barn Elms and we would talk for ages about birds and birding, and he never once bad-mouthed other birders or the birds they claimed to have found.

The first time we met was a memorable day. It was late November 1986 and I arrived at 11am, fairly late in the morning by my standards. I saw another birder slowly working his way around the filter bed. When he saw me he made his way over to me and introduced himself. Usually when you visit a site you are not familiar with it does not always result in a hearty, meaningful conversation with the first birder you meet. But Rupert and I got on famously, and struck up a great and enduring rapport. At this stage in my life I had developed a habit of continually scanning the skies and the landscape around me in search of birds. I assumed that if I was out birding then time was at a premium, as I often had to be back somewhere at a certain time. As we spoke, I looked up to see a single Bewick's Swan flying overhead, winging its way south-west through the misty heavens. For a moment I felt as though I was in Norfolk looking at a movement of waterfowl as it was hard to believe that we were watching a wild swan flying over the heart of London. It would have only left its Siberian breeding grounds a few weeks earlier. Bewick's Swans are not usually found in London and are only rarely recorded in the surrounding area.

107

We also discovered a Red-necked Grebe, another rare London bird, plus a stunning breeding-plumaged male Black Redstart. My time with Rupert was also cut tragically short, as he developed motor neurone disease that eventually claimed his life in 1992, at the ridiculously young age of 38.

Even while he was severely disabled he continued birding, and on his deathbed he was enquiring about whether a recently-found Red-flanked Bluetail, a real rarity in those days, was showing well. Humility was the chief lesson that I learned from Rupert. He always had a smile on his face, even when dealing with the occasional imbecilic birder too hung up on personal gain and competitiveness. What a man.

My final birding inspiration is thankfully still alive and kicking, and he isn't even a birder. I first met Collin Flapper on a Holyhead to Dun Laoghaire ferry in the mid-1980s. I was heading over to Dublin to visit a friend and hopefully spend some time birding at Wicklow Head, a grossly underwatched headland 30 miles to the south. The ferry trip was arduous. It was a bit stormy and grey and we were caught in a swell that meant that the boat had to wait in the middle of the Irish Sea, bobbing up and down like crazy. You don't expect to experience a rough sea crossing on a big commercial ferry, as they normally just plough through the waves as if they are on rails, so it caught a lot of the passengers off guard. I had gone to the bar to get a cup of tea but was put off by the sight and stench of the vomiting passengers, so I took refuge on deck. I was glad to be on my own away from so many ill people, and happy for a chance to catch some seabirds flying past the boat.

I thought that I would be the only person there but I noticed a young character in a beanie hat messing about intently with some music on a beatbox type of machine. I thought it was all a bit strange but suddenly my attention was averted by a low-flying Kittiwake, and as I followed it back out to sea I soon picked up a host of Gannets, Manx Shearwaters and Sandwich Terns further out. I was deep in concentration studying a sub-adult Gannet when I felt a tap on my shoulder. It was the strange guy. In a soft Irish accent he asked me what I was up to. I explained that I was seawatching and to my surprise he took a genuine interest as I pointed out the Gannets and assorted gulls that were swilling around the boat. Collin was a contemporary mix artist on his way to visit his folks in County Sligo, on the west coast of Ireland. He was working on a musical collage piece. He asked me to look after his beatbox and went down into the boat, returning five minutes later with two cups of tea.

109

Naturally, we got chatting about music and I told him that ELO were my guilty pleasure. He laughed and said that they were a good Beatles tribute band. Collin told me that he disliked most nature programmes on TV, finding them boring, but he said he would watch them if I was on. Even back then he told me that I should be telling the world about nature. I laughed.

We chatted for ages until the ferry finally made it to Ireland. After swapping numbers we kept in touch with each other and Collin, who lives in London, has remained a very close friend of mine. He has been a great person to call upon when I have needed help on my various projects as he always has an interesting and often unusual off-beat angle. He also has an insatiable sense of humour. When I worked at the British Trust for Ornithology during the late '90s he would phone the reception asking for me, announcing himself in a excitable posh Home Counties voice as 'Fllllipperty Floppity! Collin Flapper here!' The receptionist would always put him through without question.

It was when The Urban Birder was born that he really put his belief in me on the line. I had just been asked to appear on *Springwatch* in 2006 and I was nervous and uncertain as to whether I could do it. He told me that I was a natural and was born to be a TV wildlife presenter. It was his faith in me that drove me through those uncertain early days as I tried to make my name as The Urban Birder. Without that faith I would certainly have given up at the first hurdle.

110

NO COUNTRY FOR
URBAN BIRDER

To gain more knowledge, and to see the birds I had always dreamed about, meant that I had to get myself out of London on a frequent basis. I was 18 and working in Neasden Tax Office, in a job that quite frankly bored me to death and beyond. I would often go into work and not talk to anyone else, spending the whole day filing as I listened to Marvin Gaye and Prince on my Walkman. The view from my drab office window was the terraced rooftops of Harlesden, a view akin to the scene in the opening titles of *Coronation Street*.

Despite the unpromising aspect, I still amused myself watching the Swifts diving through the air during the summer, and the occasional hovering Kestrel that must have bred nearby. Once, on 1st March 1984, I was certain I saw an exceptionally early returning Cuckoo heading north at rooftop level, but I couldn't be sure. My life was split between discos, deejaying and having a bad time at home, juxtaposed with the complete serenity of birding either at Brent Reservoir, my then local patch, or the occasional weekend jaunt to East Anglia with Alan. He, on the other hand, was financially much better off than me, with a well-paid job as a trainee quantity surveyor. And he had a car, which was the crucial factor. It was an old, sky-blue Vauxhall Viva, which he kept polished to within an inch of its life. It always looked as though it had just come out of the showroom. I never worked out when he found the time to keep it so buffed up.

I well remember our first out-of-London birding trip by car, which we planned with military precision, leaving nothing to chance. It was to Minsmere RSPB Reserve in Suffolk during the first weekend of July 1983. The Great Britain road map was deployed on the floor of Alan's bedroom as we worked out the best routes in conjunction with John Gooders's *Where To Watch Birds*, opened at the Minsmere page, of course. We salivated over it as we read about the Avocets, Marsh Harriers and Red-backed Shrikes that we could expect.

These were the days before birding information technology, and what you read in books was gospel. If they painted a picture filled with ornithological delights, then that was what we expected to find when we got to our destination. The fact that the described site might be buried under concrete, or the birds that we were expecting were extremely rare, simply did not figure in our minds. Although we were aware of the existence of a grapevine brimming with the details of fairly recent sightings, we just were not in on it. All we could do was glean very basic information from some of the slightly more communicative Berkshire Bastards who we had tapped up at Staines Reservoirs.

We were so excited when our Minsmere weekend came that we decided to leave at 1am, which meant we arrived at Westleton Heath a couple of hours later. It was still dark, very dark. The sort of dark that a city boy like me was simply not used to. Alan, on the other hand, made regular trips back to darkest Ireland with his family, so took great pleasure in telling me spooky stories about wailing banshees and getting lost in fields you had previously known like the back of your hand.

113

My trepidation got worse when we pulled up outside a village cemetery near Westleton. Alan calmly suggested that we spend the night in the car, and then head to Minsmere at first light. I flatly refused, and made him drive to a nearby bit of heathland outside the village. This was near the coast and when we found a spot to park I noticed a wall of mist edging its way across the heath towards us. By this point I was freaking out, and even Alan was looking a bit concerned. We looked at each other and decided that if we were to be taken away by a UFO we would plan a mutiny together, but first we would go to sleep and worry about our fate later. We took one last look at the encroaching mist and closed our eyes.

Our first encounter with a Long-tailed Duck was not a happy memory. Luckily this male is in much better health.

We awoke just before dawn to find ourselves alive and well and still in Alan's Vauxhall. Unless we had been taken overnight by aliens and returned before daybreak after being experimented on, all was well. The birds were tweeting and we even heard the last strains of a churr from a nearby Nightjar just prior to the sun rising. It developed into a warm summer's morning, so we decided to drive to Walberswick, another area featured in John Gooders's book, and take a walk along the beach. Literally a few yards from the shore was a duck that was diving frequently. It was a female Long-tailed Duck, a lifer for us. On closer inspection we realized that it was continually diving in an attempt to free itself, due to being completely entangled in a fishing net.

We had to rescue it, but I personally hated the thought of going into the water. We were looking at the cold North Sea and I had visions of stepping into the freezing waves and being swept away to the seabed by a wicked undercurrent, for an impromptu meeting with Neptune. I snapped out of my daydream when Alan insisted that we roll up our trousers and wade out to rescue our lifer. Reluctantly, I took my shoes and socks off and, still grimacing, made my way after Alan. We used our penknives to free the frightened bird, but sadly we were too late. All the time when we thought it was diving, the poor thing was basically panicking, and by the time we freed her, she was moribund. We had to humanely put down our erstwhile lifer.

115

Our time exploring Minsmere was much more joyful, as we watched a host of species that we had previously only seen pictures of such as Yellow Wagtails, Turtle Doves and numerous Bearded Tits. The two Marsh Harriers floating over the reedbeds were new for us, as were the Spotted Redshank, Black-tailed Godwits, several Ruff with ragamuffin collars and Little Terns. The years of boning up on the birds of Britain and Europe really paid off, and seeing those species for the first time felt almost like a relief. It was as if we were collecting what was owed to us.

We also saw our first Avocets. That was a strange feeling. I remember watching a Black-tailed Godwit and seeing an Avocet standing just behind it and thinking to myself 'wow,

A first sighting of the elusive Bittern will be forever etched in a birder's memory.

there's an Avocet behind that godwit. It can wait while I process the godwit.' We didn't expect to find our first Bittern, but saw them in flight four times during the course of the day. The first time was when we were in the hide quietly trying to decipher the waders when an elderly sergeant-major type chap suddenly bellowed out 'BITTERRRRRRNNNN! BITTERRRRRRNNNN! BITTERRRRRRRRNNNNNNN!' as the bird launched itself from its reedbed hideaway. Despite shattering the hide's library-like ambience, and probably scaring off every bird within a radius of a mile or more, we were happy that the eccentric old boy had put us on to it.

Our next trip away from London did not take place until February 1984, and it was our first twitch outside of the capital. The previous day we were at Staines Reservoirs where we met a couple of visiting birders who told us of the appearance of an immature White-tailed Eagle in Buckinghamshire. The thought of a bird so huge in the middle of the countryside not too far away from us, instead of haunting some rocky northern Scottish coast, just blew our minds. We had to do the trip, so as usual we convened in Alan's war room later that day to plan our assault.

Getting to the outskirts of a town called Brill was no mean feat, especially after getting a puncture on a country lane and having to change a tyre for the first time in my life. And as if that wasn't enough we got lost several times due to my lame attempts to read the road map as Alan drove. We stood in the drizzle on the roadside by the edge of a farmer's field with a few other twitchers waiting for something we were not sure we were likely to see. Occasionally a Rook would break the horizon over half a mile away to land in the field, causing our hearts to flutter. It's quite surprising how large a Rook can look when you have no other reference points to compare it with, and if your experience of large raptors is pretty negligible.

We stood around on the same spot for nearly six hours, saw just 13 species and watched many faces come and go, until a huge set of wings cruised over the horizon with a rabbit or hare in its talons. Its immense size dwarfed the

117

attendant crows, indeed, as someone said, "look out for a flying barn door". To me it looked more like how I imagined a vulture would. I stole a look through a nearby birder's scope. Looking around and noticing other birders' modern-looking equipment made me realize that it was time to dispense with the antiques hanging around my neck and get an upgrade in the optics department. My aged Dixons pair had served me brilliantly, but equally they had reached the end of their working life. I got rid of my outdated optics for a newer model, plus a summer of deejaying helped to pay for a Kowa scope which I was to retain for the next 13 years.

We were moulting into fully fledged birders who did most of our birding in London, but lusted after birding in the rural idyll that lay outside the Big Smoke. With Alan driving we radiated out from the capital, predominantly exploring RSPB reserves in Kent and East Anglia. The next few years were spent catching up on the species we felt should already be on our lists: the sort of birds we felt embarrassed to admit that we had not seen yet.

We quietly celebrated when we clocked our first Barn Owl which conveniently flew into a barn outside Elmley, Kent, in 1984, and the Turnstones we saw at Dungeness a few weeks later in company of other more experienced birders. It was like being a Manchester United fan sitting in the Kop at Anfield, in the middle of baying Liverpool fans, when United score. You wanted to scream out loud but instead, so as not to draw too much attention, you suppress a little smile. Of course, we would not have been set upon by the birders around us had they known we were seeing a lifer, but we thought that we had an image to uphold. We supposedly knew our stuff. So we did a good job styling out a more learned approach. Don't get me wrong, we didn't pretend to know more than we actually did; we just liked acting cool.

It was also interesting being approached by other birders when we were at those country reserves. Often they would take one look at both of us and then ask Alan what he had seen. They must have thought that because I was a black guy it must have meant that Alan was teaching me, and that I was not knowledgeable enough to comment. He used to wind them up by saying I was the expert and it was he that was learning. We used to chuckle afterwards about the looks of abject confusion on some of those birders' faces. I never thought that they were being racist, it was just because they were not used to seeing a black birder out in the field.

119

Alan and I made many forays into the wider countryside during the remaining few years of our partnership, and generated many great birding memories in the process. I remember fretting over how difficult it would be to tell an adult summer-plumaged Mediterranean Gull from a similarly plumaged Black-headed Gull, until we found the former in a large flock of gulls foraging over 'The Patch' at Dungeness. We recognised it with little trouble as its totally black hood was so obvious. As birders we were very hard on ourselves to the point of feeling angry if we missed a bird flying overhead, only seeing it when it was too late to identify it. We would habitually scan the skies and landscapes looking for anything and everything, no matter where we were or who we were with.

We thought all birders were the same, but not everybody we met seemed as observant as us. Once, on a hot summer's day again at Minsmere, we were walking between hides along the beach encountering many birders who were not birding but noisily chatting to each other. This is a curious piece of behaviour that I have witnessed a million times whilst out birding. It's as though there is no wildlife to be seen when you are in a reserve until you get to a hide. On this particular occasion we watched an adult Little Gull in its smart non-breeding plumage fly at head height in full view through the groups of gassing birders, and not one of them noticed it. Absolutely no one looked up. We had the privilege of seeing an unexpected new species in superb light conditions, and to all intents and purposes enjoying it on our own.

The Force was making major inroads on many of our birding decisions. I'm a great believer in destiny and I like to think that I feel 'vibes', especially during migration periods. There were many occasions when I would call Alan to encourage him to come out birding because I had a feeling that we would see things. Invariably we would. I'm sure that

On the beach in Donegal, Ireland.

many people get hunches about good birding days for no real reason. Maybe it is just down to checking weather forecasts, or making educated guesses regarding the possibility of something interesting making an appearance. Either way, we didn't have such information to hand so instinct and a large slice of luck played a huge role.

We collectively picked up some impressive scarce species, including fine North American vagrants such as a Pied-billed Grebe in South Norwood Country Park, south London, and a Baird's Sandpiper at Farlington Marshes, Hampshire. Emanating from the east we had such gems as an Eyebrowed Thrush on Scilly, as well as the Little Whimbrel that Alan saw without me at Cley, north Norfolk. We achieved a reasonable British list with many contributions garnered during a particularly heavy twitching period that for me lasted years after Alan had ceased birding regularly. It was quite an interesting time. I always felt that I fell into twitching at the right point in my birding development. Having served a long apprenticeship and become very familiar with many of Britain's common birds, twitching added the icing to an already sweet cake.

At first it started in a very gentle fashion. In 1984, on a speculative birding weekend in Norfolk, Alan and I had met a birder in Cley who was very helpful when asked about directions to nearby Salthouse Heath. He then went above and beyond the call of duty by telling us about several other scarce birds that were in the area, plus he alerted us to the existence of Nancy's Café, which was essentially a fairly nondescript greasy Joe's situated in the heart of the tiny high street in Cley village. Inside it was the national nerve centre for birding news. No, there weren't banks of computers, loads of guys running around in white coats and thick-rimmed black glasses, or stacks of telephonists with their fingers on the pulse of the twitching world answering queries and doling out directions to all the reported rarities. Instead, it really was an average café run by a lady called Nancy who could knock out a mean breakfast.

In the corner was a constantly ringing telephone. The deal was that anyone unfortunate enough to be munching on their breakfast within arm's length of the bat phone, which sounds a whole lot better than bird phone, would be obliged to pick it up. With a mouthful of egg and bacon the gorging birder would either answer that immortal question 'what's about?' by referring to the nearby diary crammed with the latest sightings, or by using the same diary to note down the latest findings. Once we possessed that coveted number we knew that we had joined the elite. It would have been rude not to have put that number to good use.

One such call to Nancy's in early September 1984 resulted in me finding out about a Wryneck that was showing in some scrub at Halfway House along nearby Blakeney Point. The Wryneck was one of those mystical species I had always drooled over in field guides. It had an amazingly intricate plumage and, despite supposedly being a woodpecker, it looked like no woodpecker I had ever seen. Alan took no persuading, and that weekend we journeyed up to try to see it. We left at our customarily early hour and slept in the car in Cley's Eye Field car park, known as the 'Beach Hotel' by the hardy birders who travelled from all over the kingdom to have an uncomfortable sleep in their cars there. During the night you could hear the rumbling North Sea rhythmically crashing against the beach just beyond the shingle ridge that bordered the car park, interspersed with the sounds of yelping

123

gulls and the incessant 'kleeping' of invisible Oystercatchers.

At daybreak we started the long, arduous trudge along the shingle bank towards Blakeney Point itself. Each step seemed like six because our feet kept sinking into the shingle. We felt as though we weighed ten tons. After a short while we decided on a break to do a spot of seawatching. Sitting down, squinting through our scopes we were soon treated to several rafts of Eider, flocks of Common Scoter – a seaduck we usually associated with the cold winter months – and a lone Manx Shearwater that sheared and skimmed low over the waves past us. We had just resumed our trudge when I realized that I had left my bird book on the shingle where we had been sitting. I marched back and as I stooped down to pick the book up something caught my eye. I looked up and was eyeball to eyeball with a gorgeous Dotterel. We both froze and just watched each other. I called Alan over and we both delighted in watching this unusual wader, which was obviously en route to its nesting grounds on some bleak mountainside in the Cairngorms or Scandinavia. It was a first for us.

We eventually arrived at the Halfway House site, and set ourselves down on top of the bank looking down around 40 feet into a gulley at a small patch of scrubby bushes. All was quiet and we wondered if we were in the right place, but they were the only bushes that fitted the description we had been given by the Nancy's grapevine. We spent 30 stiflingly boring minutes staring intently through our telescopes at what was in effect a lifeless clump of bushes, resulting in the sighting of five flies and a bumblebee. We decided that the Wryneck must have moved on so we grudgingly decided to go back. On getting up, I nudged my scope with my knee causing it to change position slightly. I don't know what possessed me, but I decided to look through it one last time. There in my vision was a frame-filling image of a Wryneck sitting right in the open. In the sun its plumage looked more intricate and beautiful than I had ever seen depicted in any book. I instantly fell in love. Where did it come from? Where was it hiding all that time? No matter, that bit of luck still remains one of my top ten birding moments ever.

125

Sleeping in cars was something Alan and I were to do on a regular basis. It was part of our routine. We were just plain excited and always felt we needed to be on site as the sun rose so that we didn't miss a thing. Once, we nearly came unstuck when we decided to bed down in the car park of a remote country pub, somewhere in deepest Norfolk. We had arrived just before closing time and decided to have a nightcap before retiring. It was pitch black in the car park and we were settling down when I heard a noise nearby. My seat was totally flat so I slowly rose and peeped over the rim of the door to see two shady figures emerge from the pub. One guy had a torch but the other fellow seemed to holding what appeared to be a shotgun. I slumped back, and with shallow breaths woke Alan to warn him of our impending doom. We were frozen with fear as we heard them approaching the car. I suggested to Alan that we pretended to be asleep and maybe they wouldn't shoot us. He agreed and we immediately fell into a false slumber. I remember 'seeing' the bright light of the torch through my closed eyelids as I lay stiff with fear. I heard one of the men say, "it's just a couple of lads asleep". With that they walked off back into the pub. We mopped the sweat from our brow but instead of hot-tailing out of town, we went back to sleep and left before dawn.

Alan and I only ever went away on birding holidays together three times, including two successive trips to the Isles of Scilly. Our first trip was a week at Portland Bill, Dorset, at the tail-end of September and into early October 1984. It was a speculative trip, as we had heard from other birders on previous excursions about the biblical falls of migrants the Portland Bird Observatory occasionally received. We were also spellbound by the thought of the weird and wonderful things that birders had seen coming in off the sea, such as Nightjars and Short-eared Owls; and of the very real chance of discovering a rarity in the general area. That was the thing that really interested us: the chance of finding our own rarities.

As usual, I had barely any money and couldn't really afford the week, but with what I did have spare I invested in a cheap raincoat for £18 from the Macro discount score. Perhaps my best acquisition was an army parka jacket that I picked up from a second-hand market in Kensington for the princely sum of £10. What a brilliant buy. It was such a warm coat that I could have been naked under it and been able to stroll around the North Pole with confidence. To make it look a little more 'authentic' I tied it to the back of Alan's car and dragged it down the road at 30mph to give it that 'lived in look'. It was such a brilliant coat that it only went into retirement in 2006.

Our week on Portland was ornithologically poor, to say the least. We stayed in the observatory itself and during the week excursions were made to the mudflats at Ferrybridge near Chesil Beach, Lodmoor Country Park just outside Weymouth, the celebrated gull-magnet of Radipole Lake and, of course, around 'The Bill' itself. We saw very little despite searching, and by midweek we had become so despondent that we thought sleeping in was a better option. On the morning we decided to get some extra beauty sleep a Ring Ouzel showed up, the bird that I had dreamt about seeing all my life, and had still yet to see. We got out too late and missed it. We bucked our ideas up the following morning and were rewarded with our first Melodious Warbler and a rather flighty Grey Phalarope that stuck around the East Cliffs at Portland. It was especially pleasing that we found both those birds ourselves.

127

Then, on our penultimate day, the news broke that a Cream-coloured Courser had been discovered in Essex. This small, leggy desert-dweller that should have been strutting somewhere in the Sahara was a 'megatick', as birders say, and surely irresistible, being both on the mainland and so close to us. For the first time in our lives we had to make a major birding decision; do we stay and try and find our own rarity or give in and head to Essex? We chose to stay and ended up seeing nothing. Much to our dismay, the courser had spirited away by the time we had left Portland to come back to London.

Visiting all these places was a great education. I always enjoyed spending time looking at the bird that I was twitching, mindful of the thought that Peter Grant had put in my head about questioning the identification. Brief views were just not good enough and were highly disappointing. I had very stringent rules that I adhered to when it came to ticking birds. If I was less than certain of a bird's identification then it would not go on the list, despite what anyone else thought.

This notion was severely put to the test in 1988 when I was seawatching from the Burial Chamber, a high point on St Mary's on the Isles of Scilly, with a group of birders. We watched a raptor the size of a large dot hovering over St Martins, over two miles away, being mobbed by a flock of even smaller dots that were supposed to be Starlings. Simultaneously, it was radioed over as a Rough-legged Buzzard, another new bird for me, by birders standing under it on St Martin's. I could not have identified it as such from the views that I had and I would have defied anyone standing where I was to call it as a Rough-legged Buzzard. Had I seen it on my own I wouldn't have identified it, and despite categorically knowing its identity I could not count it. Such were my standards.

My love affair with twitching was beginning to wane, after going on scores of twitches where I stood waiting by a bush amongst a crowd of birding paparazzi, with hundreds of telescopes pointed in the direction of where the unfortunate rarity had last been seen. I enjoyed the banter, but was bored of the waiting; and often the end result, if there was an end result, was disappointing and fleeting. Instead of feeling exhilarated I felt as though I'd had a bad one night stand; the chase was exciting and promising but the pay-off left me feeling cheated and empty.

My two Isles of Scilly experiences in 1987 and 1988 left me with an even stronger desire to find my own birds in areas that other people didn't go to. As much as I loved the islands, I hated the crowds. People shouldn't go birding in crowds, surely? And then there were the mass illusions that I witnessed. I was lucky enough to be on the Scillies when the Philadelphia Vireo was discovered. As I mentioned previously, it was big news. I remember standing in a posse of twitchers several hundred strong waiting for this bird to show. In my American field guide it looked like a pretty little thing, almost like a warbler-shaped Blue Tit with a thick bill. Well, perhaps

129

not as bright as that, but still quite distinctive. A Chiffchaff poked its head out of the tree that we were watching and it was clearly a Chiffchaff, yet a chorus of people shouted 'there it is!' as the unassuming warbler gleaned insect prey from the foliage unaware of the panic it had caused amongst some quarters down below. Then shortly after, the real deal stepped out into the limelight. There was a collective hush and then a loudly whispered 'No, that's it there!' from the crowd. I was underwhelmed. It was a beautiful bird but that was all I felt.

Previously, in 1984, Alan and I were sitting in a hide watching our first-ever Least Sandpiper, the world's smallest wader and a bird that is usually found poking around the mudflats of the Americas. We watched it at fairly close quarters for nearly two hours, studying the differences between it and the relatively similar Little Stints that shared the same patch of mud. The American wader's yellow legs were the diagnostic feature when compared to its commoner congeners. Interestingly, there was a Little Stint present with a gammy leg, the sort of feature that you would have expected a tiny lost waif that had just flown half way around the world to sport.

A procession of twitchers came and went and alarmingly many spent just minutes looking at the bird. How could they have been so certain that they had seen it? More alarming still were the people that came into the hide and confidently ticked the limping Little Stint as the Least Sandpiper, and strolled out noisily discussing the next bird on their wanted list. I fell short of being a hardcore twitcher, traversing the length of the land at the drop of a hat, unlike quite a few of the birding friends I made during the '80s. I did enjoy visiting different parts of the country for the sake of adding another tick to my expanding British list, but I also enjoyed birding in the general area the rarity was based too. By the late '80s we had gained the level of knowledge that we had aspired to as teenagers. Alan had developed into a bit of a wader expert, being able to suss a Dunlin by jizz alone at a hundred paces. Due to the inherent deafness in my left ear I used to shun woodlands because I had great difficulty in locating where the sounds were coming from. As a result I always felt that my identification of bird calls was underdeveloped, so I over-compensated with the visual aspect of my birding. I was becoming good at noticing movement and loved being in fairly open vistas.

131

Before Alan became a lapsed birder, he and I indulged in a twitching foray in February 1989 that saw us lurking in a Tesco's car park in Maidstone, Kent, along with several thousand other birders. We were attending what is often considered to be the biggest twitch ever, looking for a Golden-winged Warbler from North America that was discovered in a nearby hedge a few days earlier by perhaps the luckiest birder on two legs. It was a male, and judging by the images and the depictions in the field guides it was a wonder to behold, with exquisite yellow 'shoulder patches' on its wings. We failed, or dipped to use twitching jargon, miserably. We didn't get within sniffing distance of it and even finding an amazing-looking Waxwing, classically sitting on a Rowan near the supermarket, was little consolation. It was also the last day of the long birding partnership that Alan and I shared. I didn't know which occasion to mourn the most.

I carried on twitching, either by myself or with an assortment of different birding mates, until the mid-1990s. Thereafter, aside from going after the birds that I thought I would never see in their natural environs, such as the Angelsey Black Lark in 2003 that had flown all the way from the steppes of Kazakhstan, twitching became less of a staple diet. In 1989 I became a media salesperson with a freshly acquired driving licence and company car. A large part of my job was to travel the length of Britain seeing clients. It was the perfect job for me as I could plan visits to parts of the country I had not been to before so that I could check out the birding sites that I had only heard of. I was content with speculatively visiting reserves or even just stopping the car next to interesting-looking habitats to search for birds.

My urban birding instincts were dulled as I had a whole wide world to explore. I was still writing the records of every species seen, heard and in some cases missed for every single expedition I made, with a supplementary notebook. It was filled with sketches and further notes on particular birds I wanted to really get to grips with. I was also still a regular visitor to East Anglia, now that I had many good friends based there, predominately in Norfolk. Places such as the pre-RSPB and scrambler-bike-frequented Cliffe Pools, Isle of Grain and Dungeness, all in Kent, were regular haunts. I made frequent journeys, usually on public holidays with a non-birding mate, to stay at his mum's caravan in Selsey Bill, Sussex. Every morning I would head to Sidlesham Ferry and walk around Pagham Harbour to see what was around. I never found any rarities there but it was always a pleasure to find Little Egrets, because in those days they were still pretty scarce and their distribution in Britain was more or less limited to the south coast.

133

So life was ticking over nicely, and my birding experiences were manifold, but I was to have two major country episodes that paradoxically were to draw me back to the bright lights of city birding.

My trusty army jacket was essential birding gear all through the '90s.

CITY 2 COUNTRY 1

It was the early '90s and Britain was going through a heavy-duty recession. By 1992 I had been made redundant after the magazine that I worked for, *TV Plus*, spectacularly crashed owing nearly £20 million. My job as a media sales executive ended abruptly when the entire workforce was called into a room for a surprise meeting with the magazine's liquidators. The next minute we were clearing our desks and heading for the exit more swiftly than a Peregrine's stoop. We were not alone as this was a sequence of events being played out across the nation at that time.

It took a little while to get another job. I applied for every position going, scoured the recruitment pages of newspapers, signed up with every agency possible and literally pounded the streets applying for any job advertised in the windows of restaurants and shops. The only benefit of unemployment was that I could spend more time birding, even while I was looking for work. Cormorants and Lesser Black-backed Gulls were the staple birds, occasionally glimpsed as they cut through the sky before being obscured by buildings as they headed for destinations unknown. If ever I wished to be a bird it was then. They've never had to worry about earning money.

On my travels looking for work I came across a job advertised in the window the now defunct Pizzaland in the Strand, central London. They were looking for a chef. Cooking has never been my forte, but at my interview with the Pizzaland head chief, a kindly Turkish guy, I talked up my ability. I don't think that he believed me, but after a few weeks on the dole I had got myself a job making pizzas. After six months of being much too generous with the pineapple helpings on the Hawaiian Pizzas I left to set up a media sales agency. It was an idea that I had been toying with for some time, a one-stop marketing and advertising sales facility for small publishers.

One of my earliest clients at Lindo Marketing Services was the fledgling *Birdwatch* magazine, which I helped to make the transition from being a subscription-only publication to a more populist magazine on the bookstands. It was while working on this magazine that I began to forge long-lasting relationships with some of the main players in the birding world, from big brands such as Swarovksi, Opticron and Nikon, through to getting to know some of the major personalities such as Duncan Macdonald of WildSounds, and my infamous long-standing mate Lee G R Evans. I was now beginning to become better known amongst the birding fraternity due to being seen birding in some of the many hot-spots and through attending the British Birdwatching Fair in Rutland, which at that time was a far smaller affair than it is today.

Two years later business at Lindo Marketing Services was looking decidedly rocky. The recession was still biting hard. I was staring skyward out of my office window one day, watching a soaring Sparrowhawk, when a the ring of a telephone shattered the peace. The female voice at the other end of the line was one that I had not heard since the late '80s. Her broad Yorkshire twang took me right back. It was Jo Thomas, the girl I had met as a teenager out birding with her dad in Norfolk, and again a few years later on the Isles of Scilly. She had pursued a career in conservation and was now the warden at Grafham Water Nature Reserve in Cambridgeshire. She had tracked me down and within a few weeks we were hanging out, enjoying many hours of birding together across Britain.

Jo was, and still is, an incredible birder and naturalist, headstrong and not afraid of speaking her mind, although she isn't a fan of cities. When I did lure her down to London she was especially adept at finding mammals. She once discovered an active Badger sett on Wimbledon Common, simply by finding a few discarded hairs in the vicinity. Badger was a species that I had never even considered as urban animal up until that point. But it was her birding skills that set her apart.

137

Her natural ability, enhanced and finely tuned by the late Peter Grant, put her in good stead. I considered myself a very lucky man to be with a birder who was far better one that me. When out, we used to work in a partnership like a couple of Harris's Hawks; she looked out for the larger birds and I for the smaller brown jobs. It was sometimes quite funny when we encountered other birders as they just didn't know what to think. They were faced with a young woman birder, a scarcity in itself, and a black male birder, an even greater rarity. Who would they go to first to ask what was around?

Jo and I had a system whereby if they asked her she would say that I was the expert and it they asked me I would direct them to her as the expert. We had a lot of fun with that plan. On one occasion a male birder came firstly to Jo, changed his mind just as he was about to speak, took one look at me and then decided to go and try his luck elsewhere!

Jo was always keen to go off looking for some wind-blown straggler, often on a whim. We twitched many birds together, but the classic time was when we journeyed to Walton-on-the-Naze, a seaside town in Essex that was the temporary home of a delectably-plumaged Red-throated Thrush. When we arrived at 8am the bird had not been seen for more than an hour. As we were both pupils of Peter Grant we knew not to stand with the crowd staring at where it had last been seen. Instead, we explored the streets and peered over garden walls in search of this beautiful eastern thrush that looked a little like a cross between a Song Thrush and a Robin. Eventually, we relocated it sitting quietly in an apple tree in someone's garden. We put the word out and were soon swamped by a posse of over 200 twitchers.

Jo also reintroduced me to Scotland on a memorable holiday in May 1995 when we wandered the Grampians picking up the Scottish specialities. She used to volunteer at Vane Farm RSPB reserve, so she certainly knew where to find the species that would ordinarily have been difficult to pin down. I got reacquainted with Ospreys and Red Grouse and had rather close encounters with a locally famous and rather pugnacious male Capercaillie in Abernethy Forest. Seeing one of these massive birds so close was a sight to behold, especially when we clocked its huge and formidable bill. This particular bird was in full display mode and took exception to our presence by charging at us with tail feathers splayed and head held high, clacking its bill. We didn't stick around to see if would actually go through with its attack and peck us to death, and instead beat a hasty retreat to the relative sanctuary of Jo's Wildlife Trust van. Or so we thought. The hefty bird proceeded to launch itself at the vehicle as we tried to get away!

At this stage of my life I was totally at home birding outside of London and on New Year's Day I would religiously make a pilgrimage to north Norfolk to get my year list off to a flying start. Jo used to accompany me, as she knew Norfolk

139

The Urban Birder's rural phase.

like the back of her hand due to her previous job at
Titchwell RSPB reserve, on the county's north-west coast.
Unfortunately, on 1 January 1995 Jo had to stay at close to
Grafham Water because she didn't have enough cover, and as
warden she had to be on site on the reserve. Initially I was a
bit disappointed, arrogantly thinking that Grafham Water
was second rate. What birds could the area give me in
comparison to visiting Mecca? Then we discovered a female
Red-breasted Merganser, a very scarce duck on the reserve.
We also saw 21 Golden Plovers fly overhead with the

hundreds of Lapwings that passed over during the day. My mood lightened. I was actually enjoying myself. In the afternoon we sloped off and visited nearby Smithy Fen to twitch a reported Rough-legged Buzzard, a scarce and beautiful wintering raptor that I had rarely ever seen. We missed that bird but had brilliant views of a hunting female Hen Harrier drifting over the reedbed. We also successfully twitched single Great Northern and Black-throated Divers, plus I saw my first Grey Partridges in years. I basically had a great day that would have stacked up well against a day in Norfolk. The lesson learnt was the very lesson I was preaching to others about; you don't have to be anywhere special to see great birds.

141

In 1997, I found myself actually living in the country. I had recently met Derek Moore, the then-chief of the Suffolk Wildlife Trust and committee member of the British Trust for Ornithology. Derek is a larger than life, gregarious man whom I had met through Jo. He told me of a position going at the BTO as its Head of Membership which was based at The Nunnery, the trust's headquarters in Thetford, Norfolk. Derek thought that I would be perfect for the job and implored me to apply. As a child I had always thought that I would end up doing something with my life that involved birds. I thought that I might have become a vet or a biologist, but I didn't really study hard enough at school. Maybe this was my calling? Well, I had to get the job first. I filled in the application form, sent it off and promptly forgot about it. Four months later I was pulling up a chair from under my desk in the Membership Office at the BTO. I had moved to Thetford and was sharing a flat with fellow BTO staffer and eventual good friend Jez Blackburn, who was then a weightlifting-loving member of the Ringing Office. I had also bought my first car, a white Ford Fiesta XR3.

Prior to joining the BTO, my impression of the organization – one that I held in common with many people – was of a crusty, fuddy-duddy, scientific and decidedly behind-the-times old boys' club. However, it wasn't as dusty as I first imagined; indeed, it seemed as though the Trust actually wanted to change its image. Well, it must have done because it had hired me. I was a bit of a fun-loving, cheeky rascal and eventually even some of the stiffer scientists who had little regard for birdwatchers warmed to me. I made it my business that they did.

My job entailed me running the membership office, liaising with the members and travelling around the country giving talks to bird clubs about the work of the Trust. I also organised events such as the Swanwick Conference, the BTO's annual gathering for members, where they could mingle and listen to lecturers speaking about elements of their work in ornithology. I knew when I joined that it would be a very different working atmosphere to that which I had been used to, and I realized that people might have thought I

142

would want to bring forth a massive seachange. But I had other ideas. For me it was all about gradual modernisation and having a laugh. There was a great social life at the BTO, especially amongst the younger guys. There always seemed to be drunken gatherings going on at somebody's house or down at the pub. During the summer there was lots of sporty action including volleyball on the lawn outside The Nunnery, football down the road, invariably against archrivals the Shadwell horse-stud boys, and badminton down at the sports centre in town. The football games were epic battles. I usually played in goal and during one game I made a superb save that would have made Peter Schmeichel proud. The only problem was that I landed awkwardly on my clenched fist, cracking a rib. I could not laugh for a month due to the acute pain.

143

Aside from the people, it was the birding opportunities that really made it for me. The Nunnery, the BTO's HQ, was right next to Thetford Forest, the largest lowland pine forest in Britain. It's a huge area maintained by the Forestry Commission, straddling Norfolk and Suffolk, that until then I never knew was originally created during the First World War to provide a strategic supply of timber. I would spend many a lunch hour walking around Mayday Farm, a section of the forest that was rich in nature trails, watching Woodlarks and looking out for Crossbills replete with their curious crossed bills. Both these birds were species that I never normally saw.

The same could also be said for the Stone Curlews that I

144

Happy days working for the BTO at The Nunnery in Thetford.

used to visit on a local common. General knowledge of these birds was kept secret, with the birding public being directed to the famous Norfolk Wildlife Trust's breeding site at nearby Weeting Heath. On balmy summer evenings I would walk deep into the deadly quiet forest, listening out for the wavering hoots produced by the Long-eared Owls that bred there.

As the light faded, and the roding Woodcocks started to patrol the forest above tree-top level, I would position myself along the edge of a ride to wait for one of my favourite sounds of the countryside: the evocative churring of Nightjars. There is something about this bird that really makes me weak at the knees. It was another of my many childhood wishes to be able to watch Nightjars flying around catching moths. I had read for years that they often showed no fear of man as they flapped around in pursuit of their prey. I would spend ages watching them do just that, and witnessing the males performing their wing-clapping display flight while uttering their 'cu-ikk' calls. I never needed to use the white hankie-waving trick to attract the birds near me. I felt like the luckiest man alive.

If I was pushed for time I would wander around the Nunnery Lakes, the then-fledgling nature reserve that the BTO itself was creating. Although I never saw anything amazing it was still nice to watch the common woodland birds that frequented the riparian habitat that led into the lakes. Some of the BTO staffers were heavy-duty, big-listing birders who would not hesitate to down tools if something rare turned up in Norfolk. The great thing was that the north Norfolk coast was less than an hour away, so it was possible to twitch stuff before or after work during the summer. Over the two years that I was at the BTO it was not unknown for me to pile into a car in a mad race to get a tick. I saw my first Great White Egret, Little Crake and Squacco Heron that way, and perhaps best of all, a superb Spectacled Warbler at Landguard Point, Suffolk. This extremely rare Mediterranean warbler looked like a small, brightly-coloured Common Whitethroat in the books, but in real life it was a different story. I wrote in

145

my notebook, 'I saw the bird after over an hour's wait. It looked like a miniature Yellow-billed Cuckoo, due to its dark ear coverts and yellowish lower mandible.'

I have a confession to make. Since a young age I have had a fear of things flying around my head. Strange as it may seem I was afraid of getting close to or even touching, let alone handling, the very things that I loved so dearly. In the early '90s, Collin Flapper found a moribund Common Gull sitting by the kerb outside his house in Maida Vale, west London. I happened to be nearby so he called me to deal with the obviously poorly bird. He waited for me to arrive, and was horrified when I took one look at it and said that I could not pick it up. I would reach down, almost hold it, then back off at the last moment. This happened a few times until, his patience lost, Collin picked the bird up and took it to work with him. He placed it on his window ledge and called the RSPCA. Unfortunately, the bird expired before they arrived.

I felt very embarrassed to be a 'bird boy' that was afraid of touching a bird. This irrational fear could be traced back to when I was a seven-year-old and cowering behind the sofa at home. Moments earlier, I was sitting on the sofa watching television. Why was I behind the sofa? It wasn't a particularly gruesome episode of *Dr Who* that got me running for cover, although some of the good Doctor's uglier mates often had that effect on me. No, it was for a different and more alarming reason. There was the clearly audible sound of

146

thousands of beating wings, deranged squawking and people screaming. I remember peering over the top of the sofa to see hundreds of gulls and crows wantonly attacking the hapless humans, pecking and scratching with a vengeance. I was watching Alfred Hitchcock's immortal movie, *The Birds*.

Like countless others, I think my inexplicable fear and dislike for anything with wings that took to flapping around my head was born right there. Anything from mosquitoes to moths to birds whizzing around my head made me feel very uncomfortable. For a small kid as I was then, that fear was magnified because the reasons for the attacks were unresolved in the film. I kept this phobia to myself until I was at the BTO. As is well known, the Trust has administered Britain's bird ringing programme since the early '30s, and one thing I needed to do was to get myself involved in the ringing sessions to get over my fears once and for all.

Apart from Jez my other best friend at work was Dawn Balmer. She was another hot birder with a penchant for gulls. Crucially, she was a ringer, so I would join her early morning ringing sessions ostensibly to learn about the wonders of ageing and sexing birds in the hand, but secretly to get the chance to hold a bird in my own hands. My opportunity came on my first morning after we had checked the nets put up near the Nunnery. After extracting a few Blue Tits and a Robin from the nets and processing them, she gave me a Blue Tit to briefly hold before releasing. My heart was pounding so much that I

147

thought it was audible, like a rugby player sprinting through fresh snow. I positioned my fingers around the sprite, as directed by Dawn. It felt so tiny and light as it trembled in my hands. I could feel its little heart beating ten to a dozen as I opened my fingers to release it. Elation was the overwhelming emotion that I felt. I had finally conquered the one thing that kept my connection to birds from being complete.

Years later I watched *The Birds* again, and admired Hitchcock's mastery of suspense; even managing to smile at some of the obviously stuffed birds swooping around without ever covering my eyes once. I was happy to have got over my mild phobia of flying things; however, even to this day the sound of multitudes of flapping wings close by still leaves me slightly uneasy.

148

Despite the incredible experiences I had at the BTO, both at work and out in the field, those halcyon days were numbered. The call of the city again haunted me. It stalked me, luring me back, sometimes for several evenings a week. I would finish in the office at 6pm then make the 90-minute drive down the M11 back to west London to hang out; only to drive back in the wee hours, have maybe four hours sleep, and be back behind my desk at 8.30am. The driving was taking its toll and I was getting progressively more tired. Things were not going so great at work either. I began to get frustrated by the lack of cosmopolitan thinking by certain people I had to deal with on a daily basis. I had always loved the ethos of the BTO, but I was beginning to feel that I was in love with something that did not love me. Maybe it was not my time to settle in the country as I felt that I had things to do elsewhere. It was time for me to leave and head back to London.

149

WORMWOOD SCRUBS:
IN FOR LIFE

Wormwood Scrubs: my inner city Fair Isle.

My love affair with my home city London was initially based on denial. For many years I was happy to get away from it, as far away as possible and as often as I could, to suck in a lungful of the bracing sea air blowing in off some remote coastline, or to smell the mix of rustic aromas produced by lavender, rape fields and silage that I had grown accustomed to while birding in the countryside. The truth is that I still love being in the middle of nowhere, but I have also accepted that I love cities, especially London. It took me a while to realise that I had an affinity with the capital. That connection made itself apparent on many occasions in my life, often at innocuous times. For instance, despite being an avid clubber, paradoxically most of the nightclubs I attended as a teenager were outside London, in fairly remote places such as Canvey Island, Essex, and Caister-on-Sea, Norfolk. I loved going to those places because even though I had entered my ebbing phase as a birder, I still loved being in locations that I knew great birds frequented.

151

London is still the place where I do the bulk of my birding. It is where I discovered the marvels of watching a local patch, and it was the place where I finally understood about what it meant to be an urban explorer. I graduated from visiting sites that were already well known and frequented by people well aware of the wildlife to be found in those popular places, to discovering an area that wasn't documented, wasn't visited and at the time, certainly wasn't fashionable.

Finding Wormwood Scrubs was to be the pinnacle of my urban exploration, and without any doubt the most fascinating place I have ever studied. My journey to find The Scrubs, as it is affectionately known, started in my back garden, then moved to Monks Park, followed by my first proper local patch at Brent Reservoir. I was on a very steep learning curve in my early days wandering around The Brent, and although I watched the site from 1975 to 2001, I was most dedicated for the first 15 years. For 11 of those years, seeing other birders there was an extreme rarity. It was either myself and Alan or later just me. Eventually, when a couple of hides were opened that overlooked the Eastern Marsh, the area that was most productive for waterfowl, more birders began showing up and sometimes I would walk around with a group of up to six birders looking for anything unusual.

Bill Oddie was an occasional visitor although I only met him there once. He was a big star but to the Brent Birders he was just a relatively local lad, being based at nearby Hampstead Heath. On the occasion that we met, Alan and I literally walked into him near the hide in the Eastern Marsh. I obviously knew who he was but didn't want him to know that so played it down by pretending that he was just another ordinary birder. However, Alan couldn't resist it and asked if Bill had been bombed by giant golden goose eggs recently – a reference to an ancient episode of *The Goodies*. I cringed, knowing full well that Bill hated being reminded about *The Goodies*. He turned and threw us a dark glare. I could see in his face that he wasn't impressed so pointing to Alan I turned to Bill and said, 'Sorry, it's his first day out.' Bill smirked and soon we were all inside a hide debating whether a Green Sandpiper we had just found feeding on the mud could possibly be an unusually dark-looking and much rarer Wood

Sandpiper. Bill and I were to become good friends in later years. Once he kindly looked after me when I visited the *Springwatch* set in Pensthorpe, Norfolk, inviting me into his Winnebago and giving me some good career advice.

I so craved to find a rarity to get my name into the annals of Brent Reservoir history. Bill did when he found a singing male Serin the following year, and several other Brent birders found a whole manner of good birds during my time there. Perhaps my biggest claim to fame was the pair of Whooper Swans in November 1990, the reservoir's first since 1966, which flew in from the west. They landed briefly in the middle of the reservoir and nervously surveyed the scene before taking off again. Birding began to get a bit competitive, with the various Brent birders vying for the biggest list. I was one of the longest-serving members but my list wasn't as good as it could have been. I was missing some of the key species such as migrant Dunlins, Curlews and the seemingly regularly wintering Firecrest. It got to the point that I would be in Norfolk watching some amazing birds, but if I heard that a species had turned up at The Brent that I needed for my Brent list you could bet your bottom dollar that I would be shooting back down the motorway to tick it. On one occasion I was watching a Garganey at Cley and then got back to London with just about enough light to recognise another skulking Garganey at The Brent, hiding behind the protruding back wheel of a discarded aquatic pram. The main

153

reason for my lower than average Brent list was that site fidelity was not high on my priorities. This was due partially to the fact that I was moving around London a bit, and also because in my heart I had not found a patch that I loved as truly being mine. I was basically birding in the well-trodden footsteps of others.

By the summer of 1985 I had moved to Sheen, near Richmond, south London. I was renting a room in an old detached house, in a wooded suburban street, that was owned by a very eccentric woman. Those summer days were filled with the sound of screaming Swifts and singing Chaffinches, and her garden was a riot of plant life with garden birds to match. I was working in Wapping in east London, selling advertising space for Mr Murdoch and *The Sunday Times*. At lunchtimes, instead of mooching around the canteen I would take a walk around the vicinity of the office. The Docklands in the mid-1980s was a mixture of dilapidated council housing estates, building sites and mega warehouse pads for the nouveaux riche. I used to observe bare-footed kids running about in the council estates and around the next block would be parked a herd of gleaming Bentleys. The dichotomy was repulsive. In a couple of the building sites I found singing Black Redstarts: the archetypal urban bird. It always amazes me to think that in the UK this beautiful bird is a complete rarity, which during the breeding season is almost completely confined to deprived urban areas. Yet just

154

across the water in Europe they are practically all over the place, like Robins are here. I felt filled with pride having uncovered my own Black Redstarts, although I was saddened at the same time because I knew that their period there was short-lived. Within a year those perfect habitats would become overpriced playgrounds for hordes of yuppies. I used to see my Black Redstarts nearly every day, just to hear the male's scratchy song and to feel the excitement within me every time I got a flash of their reddish tails.

To think that the first breeding record in London was within the Wembley Exhibition Centre building site in 1926 made my birds all the more special, because in the grand scheme of things it was not all that long ago. The Black Redstart sites were quite close to the Thames, so I would also spend a quick 15 minutes scanning the activity over the river in the shadow of Tower Bridge. I never saw anything amazing but on a couple of occasions I was lucky enough to catch small parties of Sandwich Terns, obviously following the course of the river heading into west London. These maritime birds are scarce migrants in the heart of London, so I knew I was privileged to watch them.

Back in Sheen with my nutty landlady, I had discovered that I was within 30 minutes' walking distance of Barn Elms Reservoir, a very famous birding site that was mentioned in my fabled John Gooders's *Where To Watch Birds* tome. In those days the area consisted of four basins separated by causeways

155

with a very interesting little filter bed and small pond adjoining. Of course, in years to come the site was to metamorphose beyond all recognition into the now world-famous WWT London Wetland Centre. But it was also an incredible place back then. As a member of the London Natural History Society, looking through their annual *London Bird Reports* I would read about the incredible rarities that had been seen there. The most dramatic was the account by one observer who arrived shortly after a tremendous thunderstorm one spring afternoon and was treated to the vision of an ultra-rare Gull-billed Tern that had been tracking up the Thames. It dropped in for a very short while before moving on. That was a real once in a lifetime sighting that every local patch birder dreams of: the stuff of legends.

Thames Water owned Barn Elms Reservoirs then, and you had to be a permit holder to gain access. I didn't have one, but normally the guys in the gatehouse would see the binoculars and let you through. Except on one occasion when I encountered some resistance with the question:

'You got a permit mate?'

At that point I hastily concocted a retort like Eddie Murphy in *Beverly Hills Cop II*. Pointing at my shoulder bag I quickly responded with:

'I have a very rare seabird called a Leach's Petrel in my bag and I have to release it over Reservoir No.3 so that it can pick up the right magnetic field to carry on it's migration back to the sea. I haven't got much time and, before you ask, if I open my bag to show it to you it might escape and head the wrong way. It will probably ending up dying over Oxfordshire or somewhere. You wouldn't want that would you?'

I doubt if he believed me but I think that he quite liked my cheek, nonetheless, so he let me in.

The quality of the birds to be found there was a few steps up from Brent Reservoir, although I never did see a Leach's Petrel. In three years of watching I saw interesting waders on a regular basis, with birds like Grey Plover, Little Stint and Common Sandpiper heading the list. There always seemed to be a Brent Goose or two kicking around on the causeway during the winter, while the deep basins attracted a good selection of diving ducks, grebes and the occasional diver.

157

In the summer months the air was alive with Swifts and House Martins and during passage periods Swallows and Sand Martins gathered aplenty. Sometimes skimming the water's surface amongst them were migrant Black Terns. Barn Elms was also the place I really began to get acquainted with that most glorious of our British falcons, the elegant, migratory Hobby. I once watched one individual stoop after a House Martin, catching the unfortunate creature in a mid-air upward swipe. To this day I have never witnessed another Hobby kill.

Birding at Barn Elms taught me the intricacies of identifying trickier species than the ones I normally came across at The Brent. Also, birding with more experienced birders such as the late Rupert Hastings was a godsend. Almost every visit resulted in me trying to decipher some weird hybrid duck or getting involved with the vexed subject of identifying immature gulls. I envy anyone who feels confident working out which smudgy-looking gull is assigned to which species. I could have done with such guidance when I watched a first-winter, biscuit-coloured, pale-winged gull that drifted fairly high overhead one winter's morning with a bunch of Herring Gulls. Was it an Iceland or Glaucous Gull? Both of these arctic gulls are still difficult to catch up with in the London area, and I guess my bird's true identity will be lost in the mists of time.

But it was the variety of birds to be found at Barn Elms that attracted me. I saw such oddities as Guillemots, and an errant Shore Lark that should have been scratching around on a stretch of beach on an eastern coastline, looking rather discombobulated on the grassy slope of a London reservoir. I also dipped on some national rarities including a Serin and London's only Desert Wheatear. It truly was an amazing place, where you just never knew what would turn up next, but subconsciously I was still craving a patch that I could call my own.

In 1987, two years after discovering Barn Elms, a house move to Colliers Wood, just south of Wimbledon in south London, meant that I was close to another goliath in the

London birding scene, Beddington Sewage Farm – an area steeped in birding folklore. It was a smelly sewage farm, nearly four times the size of the London Wetland Centre, and at the time you needed to be a member of a clandestine society to gain entry. The exclusion of almost all outside birders was partially to do with health and safety aspects, but mostly to do with the local birders over-zealously guarding what they perceived as being rightfully theirs. Understandably, Beddington and its birders were the bane of everybody else's existence.

Usually I would jump the fence to explore the area, and found it to be even more captivating than Barn Elms. Green Sandpipers abounded and, although I never saw one, Firecrests were regular during the winter months. The site boasted a successful breeding colony of Tree Sparrows that to this day remains one of the biggest in the UK, with some 300 youngsters fledged in 2008. I did see Curlews sometimes, a wader that was unknown to me in London, plus the unforgettable sight of 11 migrant Hobbies all sitting in a group of trees.

159

My appetite was further whetted by the knowledge that some eye-watering rarities had occurred there including, most spectacular of all, a Killdeer, which is essentially an overgrown Ringed Plover hailing from North America. This mega was found by a Beddington stalwart who watched it for five minutes in late January 1984. It briefly returned the following day to be witnessed by a few more of the Beddington regulars.

Had this bird stayed longer it would have probably meant the exclusion of the twitching fraternity at large.

The Beddington birders certainly had a reputation for being unsociable but the ones I met, including the finder of the Killdeer, treated me very well. I was even fast-tracked for a key to the kingdom. I birded at Beddington for just over a year until I moved to Ladbroke Grove, but it wasn't until I returned on a freezing cold day in February 1993 for an unsuccessful attempt to twitch a Rustic Bunting that I realised that the tide was turning. The site was under an even greater threat of development and thankfully the attitudes of the local birders had changed to be more welcoming.

I have always felt a close attachment with Beddington Farmlands, as it is now called. A very recent visit revealed that the farm is now a working site with gravel extraction, landfill and sludge spreading, and thus is still a bit smelly. It also contains flooded marshy pits that are good for passage waders, and the landfill attracts hoards of gulls sometimes with 'white-winged' scarcities and including Britain's first-ever Glaucous-winged Gull in 2007, an American west coast gull that seemingly commuted from a landfill site in Gloucester where gull enthusiasts had originally pinpointed it the previous week. The site looked very much like a work in progress and the area was clearly in need of some good habitat management. But it was still clear that Beddington Farmlands really is a sleeping giant in London's ornithological world.

With over 250 species on its list, of which more than 150 are seen annually, few sites within London's boundary can come close. I know now that access has been made easier for the general public, with several organised open days throughout the year, which is great.

On my return to west London I resumed visits to Brent Reservoir to get my urban birding patch-watching fix. I had returned more experienced and seasoned having dipped my toe on foreign soil, travelled extensively around the UK and birded some of the hottest spots in London town. The Brent was more populated with birders than ever before and I felt a stranger on a patch that I first frequented as a teenager.

161

During the summer of 1993 I started work on my commission to write the forthcoming *Where To Watch Birds In London*. It was a very exciting project for me because there had been no previous guide of its kind to the city. As I was writing about the places to bird on the western side of London I began visiting the sites, quickly finding that many of them were totally underwatched. It seemed that most birders only visited the sites with track records that indicated a chance of seeing good birds. I had to include myself in that number. Some of the places I visited were rarely quoted in the back issues of the *London Bird Report* dating back to the early '60s, yet I sensed that they had big potential. It was an incredible discovery and I wanted to adopt them all as my local patch. Some of the places were definitely birdy, with good habitat, and where the regular observers had either moved away or died. Others were exciting unexplored patches of green with no birding history, and I imagined coming back to revisit them and finding Red-backed Shrikes sitting on the tops of the bushes that also held an eclectic mix of breeding warblers.

Trawling through the records I came across Wormwood Scrubs, a place that was only mentioned twice in 50 years: a record of a 'ringtail' Hen Harrier that flew over during the winter of 1969 and a singing Wood Warbler, a scarce migrant in eastern England, that was discovered by a visiting birder in 1980. As I worked just 15 minutes walk away in Ladbroke Grove, I decided to pop in to check it out for the book. It was a warm sunny day in August 1993, and I was relatively underwhelmed by what I saw: essentially a load of football pitches encircled by a thin band of woodland dominated by sycamore, birch and plane trees.

162

Walking around I came across some paddocked scrublands and an area of grassland situated at the western edge. There was also a temporary 'tent city' occupied by several score of Eastern European students. To the north were busy railway lines and the Channel Tunnel depot, to the east and west housing, and to the south were the grounds of the infamous prison that most people think of when you mention Wormwood Scrubs. Indeed, it was sung about by The Jam in *Down In The Tube Station At Midnight* and featured in *The Italian Job*, starring the inimitable Michael Caine. It was not an appealing first impression at all, but there was something about the place. As I wandered amongst the dog walkers I noted that despite its proximity to such a dense population, the habitat at The Scrubs alerted my birding instincts to the possibility of something interesting turning up. The next few days would have read like some biblical event, almost as if the birding Gods were stepping in to change the course of my birding life.

In the past, The Scrubs used to cover a much larger area, and was part of the Great Middlesex Forest. In Saxon times and throughout the Middle Ages, the area was called Wormholt Wood or 'Snake Wood', and was used for cattle and pig farming. By the mid-18th century most of the woodland had been cut down for fuel, and it became known as Wormers Wood. At around this time the land was bought by the military and used to exercise troops and for rifle practice. During the Second World War an anti-aircraft battery and military depot were based on the site and the area used for parachute training. The existence of The Scrubs has come under threat several times during the last hundred years. At the turn of last century the area narrowly missed being turned into the first London Airport – that dubious distinction went to Heathrow! Then there was the Channel Tunnel development in the '80s. The Scrubs had survived all those pressures. It had to be a special place.

Based on my initial feeling, I decided to visit The Scrubs

163

for the next few days during lunchtime to see if my hunch was right. Besides, it was local and pleasant enough to walk around. One of the key things that I had liked about the place was the amount of sky visible. It was a surprising panorama that allowed almost unrestricted views across all points of the compass, including the Millennium Wheel, The Shard and on a good day Tower 42 in the City of London. I had learnt that looking up was a major part of birding, as so many unexpected things wing their way overhead without us ever being aware of it. The first seven days of visits resulted in no great shakes. I discovered the usual array of common breeding warblers, thrushes and finches including a fair number of nationally amber-listed Linnets and a couple of pairs of Bullfinches. On the eighth day, just as I was thinking of giving up, I flushed a Tree Pipit. This was a summer migrant that I had always associated with newly planted conifers in Scotland and the Brecks of East Anglia. It certainly was not a bird that I had seen often in London. The following day, buoyed by the previous day's discovery, I found a Pied Flycatcher, the next day a beautiful male Common Redstart, and the day after that yet another Pied Flycatcher. It was as if The Scrubs had finally yielded its secrets to me. I was hooked.

It took me a while to fall in love with The Scrubs. Initially, I used to split my time between there and Brent Reservoir, spending daybreak at the former and moving onto the latter later in the morning. That arrangement didn't last long, due to the larger numbers of warblers I starting noticing at The Scrubs, which was pretty amazing considering The Brent's far bigger areas of suitable habitat. The wooded and scrubby areas on The Scrubs ran in connecting strips and I noticed that migrants used them as a circuit route, following the tree cover from the west along the embankment area through the northern edge and then onwards east. An early morning wait by the north-west corner of the embankment during migration times soon paid dividends, with quite a few birds moving through the bushes.

The really interesting thing was seeing birds out of context. Riverine birds such as Sedge and Reed Warblers would lurk in the brambles, alongside the more expected Blackcaps, Garden Warblers and Common and Lesser Whitethroats. On the nearby grassland, parties of Whinchats would gather, perching like sentinels in the thistles. This was another very attractive species that I had previously associated with wild bracken-covered slopes, and not with urban west London. Over the years this bird became the archetypal Scrubs autumn migrant and I waited for the first arrivals with much anticipation. During my first October at The Scrubs I counted 22 Whinchats on the grassland, the largest number seen anywhere in southern England that autumn. I knew then that I had found a special place in Wormwood Scrubs. Through my study of the area I discovered that it was a very important staging post for chats in general. It was one of the

165

few places in London that you could almost guarantee seeing Northern Wheatears, Stonechats and Whinchats during the appropriate season.

I spent four years birding The Scrubs, never seeing another birder, while I built a steady list of interesting species. I watched the sky avidly and was rewarded with overflying Turtle Doves, a juvenile Cuckoo and a Hobby. By 1997, I was based predominately in Thetford working for the BTO, but I still managed to get back to London and watch my site during the weekends. But I was not alone. I was documenting my records in the *London Bird Report* and they were being read by Gary Elton, the then-volunteer warden of Hilfield Park Reservoir up in Hertfordshire. Curiosity aroused, he had started visiting my patch in the mornings before he headed to work in nearby Park Royal. He was a friendly and very communicative man, as he still is today, and a superb birder. In the two years that he covered The Scrubs he opened my eyes even wider to the notion that anything can turn up anywhere and at any time. He proved it by finding wintering Dartford Warblers for two consecutive years, an unbelievable pair of records at the time. He found a Black Redstart, discovered singing migrant Nightingales, saw a Short-eared Owl flying over, and best of all discovered a national rarity in the shape of a Richard's Pipit that he flushed from the grassland. To put that last finding into perspective, Richard's Pipits breed in Siberia and are found in the UK as an annual vagrant, usually at the established coastal migration hot-spots. They are far rarer inland and barely known in London.

Gary also inspired me to get over my last urban birding hurdle. He made me believe that every time I leave my home, Scrubs bound, I should think that I could see absolutely anything, be it a new bird for the site or a regular bird doing something I had not noticed before. A very similar lesson to the one I learnt years ago as a beginner in Monks Park. I had also become more adept at ignoring the people around me whilst I was urban birding. Buildings were cliffs and I was seeing urban habitats as how I imagined a bird would see it. A bramble patch is a bramble patch to a tired migrant whether it is located on Shetland or in a council estate in Birmingham. Your mind needs to be open to all possibilities, and that is when you start seeing things. When I say seeing things, you really do start seeing things.

167

When Gary ceased coming to The Scrubs in 1999 to move to Norfolk, I took on his mantle. Totally fired up, I sharpened up my observational skills and practised looking at everything that moved. Almost immediately I began to notice larger numbers of the scarce migrant visitors such as Common Redstarts and Spotted Flycatchers. I made meticulous notes of the birds recorded on or over the site, building a picture of the populations of the various species. It was a fascinating and rewarding study, because even the slight changes in the status of common species such as Robins were indicators of the state

Finding a Wryneck was another pivotal moment in my birding career at Wormwood Scrubs.

of the environment, giving me the impetus to start lobbying the local council who presided over the park. Luckily, the Tent City got disbanded by the mid-1990s, but the council had the annoying habit of mowing all the grassland by mid-October, just as it began to be frequented by transient flocks of Skylarks and Meadow Pipits. My requests for them not to cut the grass fell on deaf ears. Fortunately, by 2001 the regular grass-cutting ceased due to budget restrictions. A happy accident, and a great result that created untold opportunities for the grassland wildlife and flora to flourish.

By 2003 my exhaustive solo Scrubs birding began to pay off. I was walking though the paddocked area just east of the embankment locally known as Chats Paddock when a bird flew up from some cover and landed in full view on the branch of a fairly small Turkey Oak. Sensing that it was an unusual bird I raised my binoculars to take a look at it. My eyes nearly popped out of their sockets when it dawned on me that I was staring at a Wryneck. For 20 minutes I watched it fluff its feathers, preen, stretch its wings and even extend its unimaginably long tongue. Its intricate plumage was a joy to behold. I was in heaven. This was what I had waited all my life for, the chance to find a rarity on my own patch. Eventually it dropped down into some brambles and I got on the phone to put the news out. The birding world at large heard about Wormwood Scrubs for the first time.

That event would have been enough to put any site on the ornithological map. Consider my shock when just four days later I was walking along the path adjacent to the embankment when I saw what I thought was a solitary Linnet fly up and land on a wire fence. But something was wrong so I took a look at it. The bird I saw was quite robust and very shy. Not a Linnet. It had a warmish brown back with darker

169

streaks, a very evident yellowish moustache stripe on either side of its face and a whitish eye-ring. As I looked at the buffy wash across its chest, which had some fine darker streaking, I suddenly realised that it was a female or first-winter Ortolan Bunting! My heart was thumping and I was screaming inside my head. I just could not believe my own eyes. I was never alone from that day on as birders began to turn up on spec to explore my patch, and before long I had a small posse of Scrubs birders, known affectionately as the Scrubbers, who all began to find great and unexpected birds over the ensuing years.

170

We now boast an enviable list of scarcities and a decent general bird list of around 140 species that is not bad for a site without any standing water. But there was one bird on the list that meant more to me than any other species, and when I first encountered one on The Scrubs it blew my mind: the Ring Ouzel, my favourite bird. On 10th April 2006 I was sitting at home thinking about what I might come across at The Scrubs the next morning. I always get excited about going birding, regardless of where I am or where I am going to watch birds. I decided to dedicate the morning to looking for a Ring Ouzel. I had no reason to consider finding one, as there had been no remarkable movements of this enigmatic migrant thrush in the country, nor had there been any in London. I just had a feeling. A vibe. The next morning was fairly overcast and started relatively unproductively. I remember thinking that maybe I was pushing this positivity thing too far as I headed into the western end of The Scrubs. Suddenly, a dark thrush flew fairly low over my head. Like a sharp shooter, I swung my binoculars up and caught it as it landed around 100 yards away. It was a male Ring Ouzel! On my patch!

I had discovered a bird that I spent years dreaming about. I had seen my very first one rather unromantically on a twitch on Tresco, on the Isles of Scilly, in 1987. The next time I saw this species was during an intensely private moment under some bushes in the 'Desert' at Dungeness in 1989. I was quietly looking for migrants during a spring day on my own when I rounded a corner and there, literally 15 feet away from me feeding under a bush, were four Ring Ouzels. It was as if only we existed and these normally ultra-shy birds accepted my presence and allowed me to watch them with my jaw agape.

The Ring Ouzel on The Scrubs was not a one-off occurrence. I have been lucky enough to see them annually ever since that first prophetic time. It is clear that my beloved patch is on their migratory path, a fact that I would not have discovered had I not looked. How many other locations are there up and down the country and around the world that are currently under-watched, waiting for someone to come along and unlock their secrets?

My association with my beloved Wormwood Scrubs is ongoing and, I'm sure, undying. It was the place that helped me to develop into the birder that I wanted to be. It also led to me becoming The Urban Birder.

171

CALIFAUNA

Like many of us I have been greatly influenced by American culture. Americanisms have entered into my vocabulary, I love their humour and most of the people from those shores that I have met through my life have been pretty pleasant. And the music? Most of my favourite artists are American, with the man that formerly used to be Prince heading the list.

Having said that, I harbour a lot of jealousy towards America. I'm seriously jealous of their incredible bird list, some 810 species according to my Sibley Guide, the American bird bible. As a teenager I was insanely envious of the fact that they had hummingbirds, an incredible bird family specific to the Americas, of which not a single individual has ever been found anywhere near this side of the Atlantic. However, having said that, my money's on a stray Ruby-throated Hummingbird to appear here at some time in the future. With its east coast distribution and a migration route that sees it fly across the Gulf of Mexico, one of those tiny mites has got to be blown over to the UK one day. You read it here first. I was also bitter that the Americans have got a lot of 'our' birds over there too. I am not just talking about a vagrant Fieldfare on a rock on the outer reaches of Newfoundland, or the Bluethroats which just about extend their range into western Alaska. Not satisfied with European vagrants they also have their own populations of 'our' birds such as Waxwing, Golden Eagle and Winter Wren. The last species, the familiar Wren that I grew up with in the back gardens of north London, is actually the only European representative of a family that is totally American. In fact, according to current thinking our Wren is the same as their Winter Wren. So presumably they had it before us!

What is more, some of our birds are choosing to colonise The Land Of The Free. Black-headed Gulls are now breeding in Greenland and Newfoundland and wintering along the north-eastern US seaboard. Interestingly, Lesser Black-backed Gulls have been invading the States for the past 50 years, and are now apparently breeding along the north-east coast. They are frequently seen as far south as the Texas coastline yet, according to my sources, no nest has ever been found, which is quite a startling fact. They obviously are not urban birds like they are in the UK. This might surprise the good people of Gloucester, Bristol and Cardiff as they have hundreds nesting on their city centre roof-tops.

Even Collared Doves and Rose-ringed Parakeets have gained a toehold in California. I was particularly surprised to find a group of Collared Doves in the trees whilst birding at

173

The Salton Sea in the Colorado Desert. I just did not expect to see them and spent several agonizing minutes scanning through my field guide. The Collared Dove is a familiar bird in Britain and is found in most towns and cities. They are very much a part of life here, yet they were unknown prior to the early '50s, having spread from the east. The birds found in America are also largely thought to have spread there themselves, which surprised me.

Conversely, I cannot think of any American species that are colonizing the UK apart from the ones that were introduced such as Canada Geese and Ruddy Ducks, although the latter has practically been culled out of existence. There have been rumours for years surrounding the possibility of Ring-billed Gulls, one of the commonest members of its family in the eastern USA, breeding on this side of the Atlantic. To date none have been found. I even felt anger towards America when I was around nine, compiling my many lists. The reason for my angst? While writing out my extinct birds of the world list I hit upon a burning question. Why were the Passenger Pigeon, Carolina Parakeet and quite possibly Eskimo Curlew and Ivory-billed Woodpecker hunted out of existence? Why wasn't I given the chance to see those wonderful birds? To be fair, that accusation could be levelled at practically every nation on earth. I guess that it is important that we try to learn the lessons, but at times it is difficult to believe that we are.

In reality, my angst against America was borne out of a

great admiration. I was also very frustrated. All I really wanted to do was to actually see some of the wonderful birds found there for myself. Both my parents came from large families with loads of brothers and sisters who have now seemingly spread all over the world. Before my dad died in 2004 he told me that he even had family in Australia. Maybe one day I will try to track them down, but the bulk of my extended family still live in the USA.

I first visited the States as a teenager in August 1976, when I went to New York with my mum and sister. It was the first time I had been on a plane and indeed, the first time I had ever stepped outside of the UK. I remember it being a hot sticky summer in New York, and we had split our time between staying in my grandmother's suburban house in the Bronx and my uncle's rundown house in the equally rundown heart of Queens. He lived in a black neighbourhood suffering from a major cockroach infestation. It was so infested I remember going to bed absolutely knackered after killing countless of the inch-long, fat, horrible, smelly creatures. I got so fed up with it that I gave up, not caring if untold cockroaches crawled over me as I slept!

American birds were alien to me. Although I was totally clued up when it came to British birds, my knowledge of American birds was only cursory. I was only semi-familiar with some of the classic North American species such as Northern Cardinal and American Robin through seeing them

Me and my sister and cousin on a ferry cruising the Hudson. Note the stylish flares and Dixons binoculars.

illustrated in books. I was totally unfamiliar with many of the bird families and my family holiday in the Big Apple resulted in me seeing just 30 species. Most of them were ticks and most were 'page one birds'; the kinds that were very common and very obvious. The only surprise was the Black-and-white Warbler that I lucked onto treecreeping up a trunk in my grandma's back garden. I was seeing too many unidentified birds so I had to beg my uncle to buy me a bird guide. I remember him being reticent at first, taking around three weeks to get me a book, probably because he could not understand why his nephew from England was into birds. The book was useless anyway. It was bought from a local convenience store and had basic pictures that I could have drawn better with my foot, so when I saw what was almost certainly a Killdeer that I flushed from a puddle in a deserted built up area, I had to put it down as a UFO.

My follow-up visit during the Christmas of 1989 and into early 1990 was a different affair. I had come with my mate and we stayed with the same uncle, who had thankfully moved to a more respectable and cockroach-free part of Queens by this stage. It was perishingly cold, down to minus 20 with the windchill factor taken into account. I did very little birding, but a visit to a refrigerated, blustery and deserted Coney Island was memorable. I had met this girl who insisted on accompanying me. She hated me by the time we left this famous peninsula in southernmost Brooklyn. I wanted to go there because it was featured in so many films, including *The Warriors*, the cult movie shot ten years previously by Walter Hill. I loved that film, as it was an iconic call to arms for young rebels like me who really just wanted peace. Well, that was how it was interpreted in west London back then!

Aside from boring my female companion about *The Warriors*, I spent an inordinate amount of time seawatching with just my binoculars. Ignoring her moaning about the cold I managed several ticks including my first Black Ducks, Lesser Scaup and Purple Sandpipers. I also had wonderful close-up views of Red-necked Grebes, Ring-billed Gulls and a fine female Common Scoter. Years later, the scoter became a comfy armchair tick when the ornithological powers-that-be decided to 'split' the American race of Common Scoter and treat it as a separate species, Black Scoter. Needless to say, after that day I never heard from my female friend again.

Being confined to the city was not a problem for me. It was freezing with a thick carpet of snow so I didn't want to travel far anyway. I remember taking a walk through Prospect Park, Brooklyn, one afternoon and noticing a few too many shady characters. This was prior to Mayor Rudy Giuliani and his 'clean up', so many parts of New York were no-go areas overpopulated with drug-pushers and generally unfriendly people. I had a bad feeling in Prospect Park so I kept my visit short and my binoculars in a plastic bag. In fact, I seldom got them out as I walked around. Thankfully, the Fox Sparrow, a new species on my list, was fairly close to me as it rummaged on the floor of a bushy slope. I was amazed by its size; a rufous Song Thrush-sized bunting.

I fared better at Brooklyn Botanical Gardens, a venue where I felt safe to cruise around with my binoculars proudly around my neck. I saw the first of many Red-tailed Hawks, that to me looked like stocky Common Buzzards with broad short wings and short reddish tails. A Cooper's Hawk, the equivalent of a small Goshawk, headed over as I eyed up a woodpecker fest. A noisy flock of Northern Flickers passed through the area of snow-carpeted woodland where I was walking, leaving me wondering if I had ever witnessed a flock of woodpeckers before. Still pondering that proposition, I turned to see a Downy Woodpecker working its way up a trunk, and an absolutely gorgeous Red-headed Woodpecker land on the same trunk.

179

Afterwards, I made my way to the nearby Brooklyn Public Library, an impressive building with steps leading up to tall swing doors. I was standing in the beautiful atrium, about to ask the assistant for directions to the natural history section in order to corroborate the bird notes that I had taken. I then heard a commotion going on outside the building. I turned to see a young black kid being chased by several other black kids. The terrified boy was desperately sprinting up the steps to try to take refuge in the library. He made it to the swing doors but his pursuers had caught up and one of them plunged a knife into his stomach. Horrified, I saw the kid flop through the doors into the atrium as his pursuers dispersed. It was a shocking way to end a great afternoon's birding.

New York is a much better proposition now, I'm happy to say, although I do occasionally still feel nervous when I'm in certain areas. In January 2003 I made two trips to the world renowned Jamaica Bay Wildlife Refuge in Queens. It is an amazing place that covers some 9,155 acres. I had read about the multitude of birds to be found there for years, including the large flocks of wintering Snow Geese and the occasional Snowy Owls. Both were birds that I had not seen before. By now I was working as the PA to a top commercials director and had traipsed out with him to assist on a Nike shoot. It was quite a busy schedule but at the weekend I managed to steal some time to check out Jamaica Bay. I was staying in an apartment in downtown Manhattan and despite having a late night at a bar with the crew on the Saturday I excitedly returned to the apartment to plan my route using the subway. The great thing about NYC is the fact that it really doesn't sleep. The trains run all night as do some of the clubs and diners.

I left the warm sanctuary of my apartment at 4.30am while it was still dark, and started my subway journey. For the first time I had brought my scope and tripod with me to the

States, but I took the precaution of bringing a knapsack to house my optics out of sight. As I headed into deepest Queens the subway clientele changed markedly from shy Latin cleaners to dodgy-looking hoodlums who had the perturbing habit of walking through the carriages as if they were casing the joint. I tried to style it out, pretending I was a native New Yorker and that this was just another day. A couple of guys did sit near me and looked quizzically at my tripod stand, but that was the nearest that I got to any drama.

My visits to the refuge were incredible. It felt like being in the wilderness, yet if I glanced over my shoulder I could see the famous New York City skyline across the Hudson River, jutting out from the horizon like a set of uneven metallic teeth. I got to see over 300 Snow Geese roosting on the mudflats as well as more than 1,000 Black Ducks, the most common bird I saw. The males and females were similarly plumaged, looking like dark chocolate Mallards with silvery-white underwing coverts. I didn't see a Snowy Owl, and I think that their regularity there has probably been exaggerated, but I did flush a magnificent Goshawk from fairly close quarters. Jamaica Bay is the sort of place you need to visit regularly, but unfortunately I have not yet had the opportunity to return.

The reason for my disappearance from the New York scene was that since 1998 I have been at least an annual visitor to Los Angeles. I skipped the middle bit of North America and landed in Tinseltown. Well, West Hollywood to be exact; for sun, sea and birds, oh, and work too. I had followed my director boss to his new venue for making commercials. Los Angeles is the City of Angels. Surprisingly for me, it was also a city stuffed full of birds. I certainly did not expect this, based on many of the films I have seen over the years. It is the second largest city in the States, a massive sprawling place with a population in excess of 14 million. A true 'megacity' in global terms.

The classic image most people have of LA is of great weather, big flash cars, beautiful (and often physically augmented) people, most of whom are wannabe actors – and, of course, Hollywood stars. Those things certainly exist and I killed many an hour just observing that side of LA life. The metropolis is made up of a collection of different cities including my base, West Hollywood; a place whose salubrious streets and its surrounding areas such as Beverly Hills and coastal Santa Monica were to become my periodic stomping ground for over ten years.

It's true to say that nobody walks very far in LA. Practically everyone drives everywhere, including me, as I usually had access to a car on my visits. If you were to take a stroll in West Hollywood in the dazzling sunshine you would notice that even the Feral Pigeons seem more lethargic than our birds in Britain, as they loaf around on top of the massive billboards. On my initial strolls I saw plentiful American Crows and

Ravens swooping over the streets. The Ravens in California seem smaller than the wild montane birds I was used to and there is talk amongst the scientific community of splitting them into a separate species. I would certainly buy into that notion. Western Gulls flying around over the city were also a common sight. These birds were like broad-bodied, pot-bellied Lesser Black-backeds with shorter wings. Mourning Doves were ubiquitous, and while standing on the roof terrace of the swanky hotel where I was staying I occasionally caught sight of the much shyer and scarcer Band-tailed Pigeon, a bird with the jizz of a slim Woodpigeon.

It was the hotel rooftop that became my first patch in LA. In the early days I was on a short leash, due to being on call most of the time. From my rooftop haven I had good views of the Hollywood Hills and distant downtown LA to the south-east, with its cluster of tall buildings that were always shrouded in the orangey early morning smog. On the horizon was the edge of the city and the sea. Yes, Los Angeles is not only a desert city but a seaside city too. I chose my times to be birding from the roof carefully. First thing in the morning was great, but it meant there were few raptors or hummingbirds around, as they were still warming themselves up. By late morning there was a lot of hummer activity but my domain was now shared by Hollywood luvvies discussing scripts, either soaking in the jacuzzi or knocking out a few laps in the small pool. Curiously, very

183

few people sunbathed on the roof. It was at that time of day that I would often see a Red-tailed Hawk distantly launch itself into the air and lazily ride the thermals above the unsuspecting city.

Checking the treetops around the hotel often revealed hummingbirds busying themselves amongst the blossoms, in the company of Audubon's Warblers, the default wood warbler here. I did once chance upon a gorgeous migrant male Wilson's Warbler in the foliage and Western Scrub Jays were not unknown from my lofty perch either.

184

Some years I used to visit the city up to three times, sometimes staying for a month at a time, so naturally I began to explore. As I mentioned earlier, West Hollywood is bordered to the north by the Hollywood Hills. This area serves as a sanctuary for many of the movie world's glitterati and is strewn with big houses, many of which look pretty bizarre. It was also the location for one of my birding stomping grounds, Franklin Canyon, which was only a 15-minute drive from the hotel. This 'patch' was essentially a fairly large wooded valley encircled by a road and with a small natural-looking reservoir at the southern end. I would see all sorts of birds there, including Acorn Woodpeckers with calls sounding just like Woody Woodpecker, the cartoon character. One summer afternoon I was birding from the road, watching a singing Song Sparrow, when I heard a car horn behind me. It came from an open-topped Bentley driving slowly past. The rotund bubbly smiling face behind the wheel belonged to none other than chat show host Jay Leno!

I used to play football on a Beverly Hills baseball pitch down the road from the hotel. Even while throwing myself around I was constantly on the lookout for birds. Brewster's Blackbirds, a member of the icterid family, were commonplace, as were the American Crows, but I had many special moments from this very urban setting. Among my favourites were the regular winter flocks of Cedar Waxwings that used to swoop past to raid the bounty of berries in the hills. There was also the time that I saw a tired, solitary Western Meadowlark sitting on the pitch one morning. This is a curious, yellow-and-black bird, shaped like a Starling but with a flight like a Green Woodpecker. One early evening in the same park, I was lying on my back staring into the sky, absolutely knackered after a marathon football session, when I noticed six American White Pelicans heading over, illuminated by the floodlights.

On one of my urban birding wanderings, I met Dick Barth, a retired casting director and cameraman who had worked on some iconic Clint Eastwood films, and who also turned out to be an amazing urban birder. He was incredibly observant and could hear a bird tweet a mile away despite being hard of hearing. I never did quite understand that one. What I loved about him was that he went birding anywhere from a small park in the middle of busy Sunset Boulevard to

185

'proper' birding spots. But even then he went to places that the crowd didn't. He took me to a small park on Sunset for a twitch. The patch of green was akin to a roundabout island in Britain and contained just four palm trees. Sitting stock still on one of the trunks was a vagrant juvenile Red-naped Sapsucker. Although it was surrounded by tiny holes that it had previously excavated it did not move an inch during the whole 20 minutes that I was watching it. Most un-woodpecker like.

Dick once took me to a tiny and very scary park in downtown LA that only contained about ten trees. He had found a wintering Black-and-White Warbler, a rarity on the west coast, a few days previously, and had brought me there to see it. The problem was that some very, very shady characters also inhabited this park, a couple of whom were openly scoring drugs yards from where we stood. Dick could not care less as I watched the pied sprite through trembling binoculars. He was a true urban birding legend who would explore anywhere and find anything. That is how all birders, naturalists and wildlife enthusiasts should be.

Travelling around the city checking out potential birding spots was a lot of fun, but as ever I had a gnawing urge to study one main patch that I could collate records for and really get under its skin. That place was Ballona Wetlands. This remarkable and beautiful place, pronounced 'Bi-yowna' is situated on the coast between the city's main airport and

CALIFAUNA

Marina Del Rey, an arty town a bit like a mini Notting Hill by the sea. It was an easy 30 minutes drive from the swanky streets where I was based. The site is a 1,087-acre ecosystem consisting of saltwater and freshwater wetlands, dunes and upland habitat. The Ballona Wetlands that exists today is a result of years of fighting. As with most urban nature reserves there have been tremendous pressures from the threat of development. For over 100 years the site has had a truly turbulent existence, with Hollywood-style drama thrown in for good measure. Howard Hughes bought over half the land in the '30s and more recently Steven Spielberg was only just prevented from developing the area into a studio.

187

I first heard about its struggles during the late '80s, and felt compelled to help the cause. I joined the Friends of Ballona Wetlands, and later signed online petitions. By the time I made it there for the first time in 2000, things were looking a lot more stable. I pulled up in the small parking lot next to the freshwater pool and before I could even get out of the car the Allen's Hummingbirds darting around the blossoms on the edge of the lot caught my attention. The small freshwater pond had great-looking muddy fringes that supported both Snowy and Great Egrets, while the two waders I picked out on the mud itself included my first proper Killdeer and my first-ever Greater Yellowlegs. Swimming on the pond's surface were a couple of Pied-billed Grebes and a few Mallards that were consorting with a group of Lesser Scaup and Buffleheads like domestic ducks on a village pond. I had hit the jackpot.

I kept on walking around the edge of the pond and bumped into a group of birders on a field trip. Their leader was Bob Shanman, a very amiable guy who immediately invited me to join them. We moved onto the adjacent beach area where we sifted through the load of assorted gulls and terns that were resting literally yards from where we had congregated. We sorted out the California Gulls from the Ring-billed, Heermann's, Western and American Herring Gulls; plus there were Elegant, Royal, Caspian and Forster's Terns to add to the mix. It really was incredible to be so close to birds that I thought would have been frightened by the *Baywatch* scene that was being conducted all around them. By that I meant the sassy girls on rollerblades speeding past, cyclists and walkers passing along the concrete cycle path that split the beach between the birds and us. There was quite a funny moment a few minutes later when we walked away to look at the Brown Pelicans and Pelagic Cormorants assembled on the nearby rocky jetty through which the Ballona Creek flows. I noticed a Caspian Tern fly past so I pointed it out to the group explaining that they were to be found all over the world.

"No they're not! They're American!" exclaimed an elderly lady birder in the group.

"Err, they have a worldwide distribution…" I tried to explain before being being cut off once again.

"No! They are American birds!" she retorted.

Sensing an impending breakdown in diplomatic relations, plus seeing the other members of the group suppressing smirks, I raised my hands in submission. Bob seemed impressed by the way I handled his patriotic group member and asked me to lead a tour the following morning in Malibu Creek State Park. So the next day I was taking 20 American birders around showing them birds. I would see a bird and think 'Wow, that's a Nuttall's Woodpecker! A tick!' and then calmly turn to the group and nonchalantly point and mention, 'Oh, there's a Nuttall's Woodpecker.' While they were enjoying that bird, I was already looking at another lifer, quietly celebrating before putting the group onto it. They thought that I was great!

189

Thus began the annual visits to my LA local patch. Because there are three sets of habitats at Ballona – salt marsh, freshwater lagoon and the sea – over a period of time the pattern of birding begins to take shape. I learnt that certain times of the year were great for waders, ducks and seabirds, and with the help of the more experienced local birders I became more comfortable in identifying the trickier species. It wasn't always a blazing sunny day either, some mornings during the winter could start quite chilly and very misty and

190

America's Hudsonian Whimbrels seem much more confiding than their Eurasian counterparts.

I felt as if I was on the Norfolk coast looking for migrants. I don't know what it is about America, but for some reason you can get a lot closer to some species than you can in the UK. For example, I have walked right up to Whimbrels on the beach at Santa Monica whereas I have not got within 300 feet of a Whimbrel at home. Birding in LA was a pleasurable surprise in terms of the number of species and the number of sites to watch them in. It proved that the principles of urban birding can be applied to almost any city, anywhere in the world. All you need to do is look. To this day Bob and I are great friends and we still keep in regular contact. He recently told me that Dick Barth, who must be rapidly approaching his 80th year, is still out birding every single day, sending in daily records to the city's birdline from the strangest corners of Los Angeles. That's what I want to be doing, wherever I end up.

191

THE WORLD TOUR

My six years as PA to a top commercials director, between 1998 and 2004, meant that I travelled frequently – and not just to LA. I had landed on my feet as the director was a great friend of mine, very well respected in his field and crucially, he needed an assistant. It may have sounded cushy, but it was often far from it with very long hours that involved being stuck in buildings or in an indoor set far away from the nearest bird. The locations I visited always sounded fabulous but that didn't mean that I got to see them.

Take for instance the time I was on a job in Cape Town, South Africa, in 2001. On that occasion I was working such long hours that on my down days I was literally too tired to get out of bed. On my final day I dragged myself out of my pit after barely an hour's sleep to jump into a cab and visit Rondevlei Nature Reserve, an urban birding hot-spot. I only had a couple of hours there as I had to get to the airport. On arrival, I clambered bleary-eyed out of the cab to discover that I had left my binoculars back at the hotel. Most of the filming locations were in cities so it was often a great challenge for me to steal some time for a quick birding session at a nearby venue or even to indulge in a little urban birding on the job.

A lot of the commercials that I worked on were UK-based with locations as varied as a Soho strip club to a downtrodden car park adjacent to Newcastle United's ground at St James' Park, where we were filming a commercial for Barclays Bank with the late, great Sir Bobby Robson. He was a fantastic down-to-earth guy and when not idly chatting with him I was sneaking off to listen to a migrant Willow Warbler sing from a patch of scrubby brambles and Buddleia. I was actually quite surprised to hear any bird singing as there was barely any cover at all.

193

We also shot an advert in Mold, north Wales, with footballer Michael Owen. It was a mid-January shoot and brass monkeys. I spent a lot of time running between Michael's home, where we were filming, and the unit base. The local residents took pride in their provision of food for the birds and my regular jaunt took me past some well-stocked gardens. My diligent peering over the garden hedge was rewarded with views of some lovely Siskins gorging on someone's seed feeders, and an unexpected wintering male Blackcap.

Travelling abroad was an interesting proposition. From a birding prospective it was both frustrating and challenging. The frustration was borne out of the limited time and potential birding venues that I had available to me, but the

A male Blackcap is on the very edge of its winter range in north Wales.

exciting challenge was to make the best out of the situations I found myself in. A spring trip to Malaga for a beer shoot entailed a fortnight based at a plush hotel outside of town, just across the road from a building site. Apart from birding from my balcony in my downtime, resulting in a fortuitous sighting of a migrant Great Spotted Cuckoo flying in off the nearby Mediterranean Sea, I made daily visits crossing a treacherous road to my adopted building site patch. Even though I saw no more than 20 species it felt as though I was there to personally welcome all the individual birds that passed through. During the first few days I found Nightingales and Blackcaps while the few Serins and Sardinian Warblers that I saw appeared to be residents. By the tenth day I remember feeling elation when a flock of 12 Bee-eaters arrived to spend the next few days catching bees from the telegraph wires that traversed the building site. They were my own private birds that had chosen to grace my little patch which no one else in the world would have thought to cover.

When stuck in a city, even for just a few days, I would always try to discover a little spot to visit, find birds and ground myself before a busy day. Perhaps my biggest challenge to date was when I was based in a hotel in a densely-populated inner city, red light district of Mexico City. We were shooting an HSBC Bank commercial that would keep us there for two weeks. Having checked into the hotel with the team the next thing I did was to discreetly check were the nearest green spaces were. My room was pretty high up and the view from my window afforded little sky and instead, mostly the drab sides of other similarly tall, boring buildings. Outside the hotel the few trees that lined the noisy polluted streets held tiny finch-like Inca Doves, House Finches and House Sparrows. The nearest park was a small, lightly wooded affair in the grounds of a couple of museums, some ten sweaty minutes' walk from the hotel. I use the term 'lightly wooded' in the loosest possible way, as there was no undergrowth

195

whatsoever, and the park was invariably filled with people walking their dogs, exercising, or sleeping off excess drink. Having been warned about the high risk of foreigners being mugged, kidnapped or worse, I tried to blend in as much as possible by doing the 'hiding my binoculars in a plastic bag trick'. The problem was that the only bag I had at my disposal had 'Apple' plastered all over it. It may have well have said, 'Hey guys, check this out! Steal this!'

I visited my edgy patch every day at varying times, and to my complete surprise it turned out to be a migrant treasure trove, with a host of colourful transient visitors on their way to North America. Every day seemed to bring in a new set of birds, and while watching a newly arrived Ovenbird, an American wood warbler that reminded me of a cross between a Robin, Dunnock and Goldcrest as it gleaned insects from the ground, I found a thrush feeding in the undergrowth that I was unfamiliar with. Looking in my Mexico bird guide I worked out that it was a Rufous-backed Thrush, a bird similar in size and shape to our familiar Blackbird. But according to my book this attractive thrush did not have a range anywhere near the city, yet there I was staring at one. Eventually, I realised that some of the migrant American Robins that I had seen previously during the week were in fact Rufous-backed Thrushes. The Mexican thrush was shyer and more retiring compared to its American cousin and that was why I had overlooked them.

A few days later I was working in another very busy part of the city. At lunchtime I walked into a public park with large, evenly-spaced trees, and small lawns with cute little decorative metal fences around them. There, hopping on the lawns, were Rufous-backed Thrushes totally unfazed by the human activity around them. It was quite interesting studying them as they had a more fluttery and undulating flight, often ending in an upward swoop to perch within a tree. The American Robin's flight was slightly more direct and purposeful and it often landed out in the open. It was all the proof I needed that urban birding in the parts of cities deemed least likely for birds should be the first places to be explored. I love the thought of being an explorer where no one had been before, yet right in the middle of a big city.

197

Not all my trips abroad were necessarily fruitful for birding, and could never compete with the birding possibilities that a holiday could offer. Three of my most frustrating birding sessions whilst working abroad all involved Guinness. My first proper work trip abroad was to a beautiful town called Monopoli in the Puglia region, right in the heel of Italy. It was an ancient town populated with some very friendly people. I initially came out with the director and the creative team to recce locations for a few days prior to shooting a Guinness commercial that featured an aged swimmer. We stayed in a grand country hotel in Fasano, north of Monopoli. It was April 1998 but, far from being sunny and warm as you would expect in southern Italy, we were treated to snow showers that even raised eyebrows amongst the locals.

One evening I was sitting in my room when I got a call from the director requesting my help in removing a bird trapped in their meeting room, which had got in via the fireplace. Now, despite saying that I was 'cured' of my predilection for running in the opposite direction whenever I was called upon to handle a bird, nervousness set in. This was a case where as 'The Birdman' I was expected to know what to do when it came to capturing birds.

I opened the door to the meeting room and walked into a Tardis. The guys were in a massive ornate room with a mile-high ceiling, the aforementioned fireplace, hanging bling-bling chandeliers and an en-suite bathroom. From the corner of my eye I noticed a male Black Redstart flutter up to land within one of the chandeliers. My lips quivered with trepidation. What followed would not have looked out of place in a Carry On film. Grown men sitting on top of each other's shoulders, trying to shoo the bird out of its hiding place, then placing a table under it, then a chair on the table. Then one of the guys stood on the chair and suggested that I got on his shoulders. Aside from looking like a bad audition for Cirque du Soleil I also didn't want to end up in casualty. Before I could open my mouth, one of the creatives had a very bright idea. He suggested turning off the lights in the main room leaving the bathroom, with its much lower ceiling, illuminated. His thinking was that, like a moth, the bird would fly to the light in the bathroom, rendering it easy for capture – by me. Great plan. Right.

The lights went out and, as predicted, the bird flew into the bathroom. I was dispatched to deal with the bird. Sitting in the corner of the ceramic sink unit was a very scared-looking Black Redstart. We stood facing each other. We must have been locked in a gaze for some time because the same creative came in to see why I was taking so long. Immediately reading the situation he grabbed the hand towel I was feebly holding and quickly threw it over the bird. He gently picked up the bundle and handed it to me. He smiled at me, and as I walked out of the room with Black Redstart in hand he announced that I had caught it. Embarrassing.

It was Hawaii, no less, that was my next assignment. I remember being so excited about the prospect of visiting a Pacific island and watching lots of birds that most people never get to see. We were shooting the now world-famous and much awarded Guinness 'Surfer' commercial, featuring white horses morphing out of the waves. We arrived on the island of O'ahu at nightfall, after getting a connecting flight from LA. I never realised just how far Hawaii was from the rest of the States. It was a six-hour flight. Nor did I realise just how big O'ahu was. It took us nearly two hours to drive from Honolulu in the south to Kahuku, the northernmost point.

199

Once ensconced in a hotel that overlooked the sea and rocky coastline, I spent the remainder of the night sitting on my balcony waiting for dawn, and wondering why there were such big waves crashing onto the rocks but with no accompanying wind. When dawn finally came I was down on the rocks for some pre-breakfast and pre-work seawatching. Within minutes I was watching a breaching Humpback Whale on the horizon and had found a Wandering Tattler working its way around the rocks just yards from where I was sitting. It looked like a grey, unstreaked Redshank that kept on bobbing on short yellow legs. Brilliant, I thought. If things carry on like this I would have an amazing trip.

I spent the next two weeks standing topless on various beaches watching a helicopter filming guys surfing, which resulted in me getting sunburnt shoulders. My half-nakedness was due to the heat and the fact that I still had a reasonable body that I didn't mind flaunting. But I never knew that I could suffer from sunburn. The birding for the rest of the time I was there did not live up to that first morning's promise. Despite seeing my first-ever albatross, a Laysan Albatross that looked like a rangy-winged Great Black-backed Gull, plus abundant Pacific Golden Plovers that frequented roadsides, fields, golf courses, car parks and even congregated on rooftops, I saw a total of just 13 species. What I had not bargained for was the fact that Hawaii was one big open-air aviary. Liberated birds emanating from all corners of the earth made up the bulk of the species that I saw. The endemic passerines I had hoped to find had been all but wiped out by the avian malaria brought in by the introduced birds, and they now only existed at altitudes above 4,000 feet in the mountains. Thanks to our coastal shoot locations I never got the chance to look for them.

A few months later I was on yet another Guinness commercial, this time in Cuba. The advert involved a bunch of racing snails and again I was beside myself with excitement when I considered the quality of birds I could be watching while I worked. For the first few days I was in heaven. Our set was based in the jungle, around two hours' drive from Havana, and my plan was to set some time aside to go birding around the set every lunchtime. It was all going to plan until the third day when a couple of Cuban colleagues warned us that a hurricane was heading our way and that we needed to retreat to the hotel in Havana for a few days. They were not wrong and the next day Hurricane Irene hit with force. Our hotel was not too far from the sea, so it was incredible to watch the rain falling horizontally and the sea lapping over the coastal roads with a ferocity that I had never seen before. It was genuinely frightening.

After four days of being cooped up in a weather-battered hotel eating chicken and chips for dinner everyday, the storm finally subsided. The damage on the streets was devastating but worse still was the news that our country location, my

201

ticket to birding paradise, had been flattened. A new location for the shoot had to be scouted and when I heard where it was the news could not have been worse: it was in a Havana multi-storey car park, and not even on the roof. My birding opportunities diminished to nothing and during the shoot the only thing with wings I noticed was a massive cockroach that landed noisily on a wall next to my head with a sickening, crunching splat. Yuk!

I could not recollect some of my urban birding escapades in the name of advertising without mentioning birding in Carnarvon – not the place in Wales, but the dusty two-horse town in the Karoo, South Africa, in 2002. We were filming a music video for the band Travis, and I was equally excited about hanging out with the band and birding in this brilliant part of the world. When we rolled into town it was immediately obvious that apartheid was alive and kicking. The blacks lived in the shabby side of the settlement and unbeknown to us, the hotel that we were booked into was effectively a whites-only establishment.

When I walked in it was like a scene from one of those movies: the whole bar went quiet, heads swivelled and all eyes were on me. In the mornings before dawn I would venture out on to the dusty streets looking for birds. Boy did I find them, in abundance. On two separate mornings I also came across two white women, one walking towards me with her dog, the other jogging. On both occasions I bid them good

One of the most delicate falcons, a flock of Lesser Kestrels was one of my abiding memories of South Africa.

morning, and on both occasions the women shot me a look as they silently crossed the street to carry on their journeys. I felt saddened. Not for me but for them. What must they be missing out on in their lives?

My favourite birding moment in Carnarvon was after spying a flying Ludwig's Bustard. I was celebrating and feeling great when a group of ten gorgeous Lesser Kestrels suddenly appeared. I was immediately overwhelmed by their sheer beauty and grace as they silently hawked for insects. I was entranced by their light, flappy flight, just a few metres above my head, against the backdrop of a beautiful deep blue sky. I stood there and watched them for what seemed like an eternity. Time stood still. It was just them and me, man and birds. Our rendezvous ended as quickly as it had started when they slowly drifted off into the bright sky.

204

BIRDS ON FILM

'Have you ever seen a psychic?'

This is not a question that you get asked every day, but it was asked of me once by a Production Manager colleague on a set in Pinewood. I had to stop and ponder. The thought of seeing a psychic was both intriguing and nerve-wracking. How was I to know that a couple of months later I would be sitting in the front room of a house in Watford, eyeball-to-eyeball with an actual soothsayer? I would be there as the result of my colleague taking it upon herself to secretly book me for a consultation to see 'Sue' the psychic.

I remember turning up outside the psychic's house in May 2005, feeling a little uneasy as the evening sun beamed brightly upon a crowd of screaming Swifts which were swooping low over the rooftops in full chase mode. Once inside and seated she started breaking down my love life. Sue was a very warm and friendly woman with a nice energy about her. She laid some tarot cards on the table and suddenly her expression changed. After re-dealing the tarot cards and studying them for a few more moments a broad smile spread across her face.

"This is amazing! You have a fantastic life ahead of you! This is great – this is so exciting! You are going to be very successful. You will be doing something you always should have been doing. Isn't that exciting?"

She looked at me as if I should have known what she was talking about. I didn't have a clue. I stared back at her blankly. What was I supposed to say?

"Ohh! Look at all the places that you are going to visit! I can see you in a desert area and can see loads of those wind turbines behind you. And there you are in Kentucky – well at least, there is a sign saying Kentucky here."

It felt like she was describing Michael Palin's *Around The World In 80 Days*.

"I can see you in New Orleans. You're in a grand building, there's music playing in the background and you look like you're taking some notes. You look so happy. Make sure that you come back to see me with pictures of these places so that I can see exactly where you were."

My scepticism started to kick in again. I mean, she never answered the crucial question: would I be the person who would rediscover the almost certainly extinct and totally enigmatic Eskimo Curlew?

The session ended shortly afterwards. After bidding Sue farewell I strolled out of the house to my car which was parked nearby. The sun was setting behind the suburban

206

landscape of north Watford as a few groups of Swifts still swirled between the houses, screaming as they continued to chase each other in the fading evening light. I felt strangely lifted and drained at the same time. Confused. I didn't understand why, as she hadn't said a lot about anything that I could get a handle on. My hand was scooped deep in my coat pocket cradling the audio cassette that she gave me as a memento of my visit. I got into my car, thoughtful. Within an hour I had relegated my Watford psychic experience to the back of my mind. Six months later, in January 2006, I received an email from the BBC Natural History Unit in Bristol – totally out of the blue. It was an invitation to be filmed birding at Wormwood Scrubs for *Springwatch*. My jaw fell open. Where did that come from?

The shoot was planned for April 2006 and the weeks and months just flew by. The night before the shoot I met the production team in their far-from-five-star hotel in Acton, for dinner and a briefing for the shoot day. The director put me at ease by encouraging me to be myself on the day. She also asked if they could give me a screen test afterwards. This last question excited me. Why did they want to give me a screen test? Trying to curb my outward enthusiasm I agreed to her request as casually as I could. Back at home I got on the phone and spoke to a couple of good friends for advice. They filled me with confidence by telling me they thought that I would be great on screen, that I should be myself and that I should wear the cool clothes I normally wore and not the typical bad wardrobe choices displayed by many nature presenters.

When I got home I paced up and down the kitchen wondering what I could say or do that could potentially make me stand out as a possible new wildlife presenter. I figured that I would play to my strengths. Like everybody on the planet interested in nature, I too wanted to be a David Attenborough, wandering the wilds of the world stroking penguins and having close encounters with Tasmanian Devils. But I also knew that I had a thing with city wildlife and that no one in the public eye seemed to be championing urban conservation in the way I wanted to. Globally. To my mind, urban wildlife had always been treated as a novelty, a poor cousin, or as something that happens by accident, and I felt quite strongly that I would try to represent that cause. Influenced and inspired by Jamie Oliver's *Naked Chef* I set about thinking of a comparable moniker. City Birder? Town Twitcher? Then it suddenly came to me. The Urban Birder? Yeah, The Urban Birder. It encapsulated all that I wanted to be about and it had funky ring to it.

The next day we started shooting on The Scrubs at 4am and kept going until around 3pm. It was quite a flat, grey day and I was freezing cold and tired throughout, but had an incredible time. I remember thinking that if it all ended here, on the first day at the first hurdle, then my life would have

208

been worth living. I called myself The Urban Birder during one piece to camera, and it made the final cut that appeared on *Springwatch* in June 2006. I was in Los Angeles at the time of its initial airing and my Blackberry was buzzing with loads of text messages and emails from everyone ranging from my mum to people in TV, saying how much they had enjoyed my performance. I felt happy and I couldn't help but think back to Sue the psychic's words.

I wondered how the BBC had first got to hear about me. The answer came a couple years later during a meeting I was having with Tim Scoones, the executive producer of *Springwatch* at the NHU in Bristol. At the end of the meeting he casually mentioned that he wanted to introduce me to producer Kirstine Davidson, the woman who first discovered me. As Tim dialled her extension I was wondering whom would I be meeting? Moments later Kirstine walked into the room. She was far younger than I imagined. After getting all the pleasantries over with she broke down the chance sequence of events that led to me first appearing on *Springwatch*.

In 1998, whilst I was Head of Membership at the British Trust for Ornithology, I appeared in a Reader's Digest video about birding narrated by the great Tony Soper. I featured in a five-minute segment wandering around Brent Reservoir and Wormwood Scrubs talking about urban birding. In those days I used to have dyed blond hair, so I certainly stood

209

out, but it was such an insignificant piece that I wondered if anyone would ever see it. After shooting it, I promptly forgot about it. Cut to the early noughties and an even younger Kirstine, then a researcher at the BBC, was looking after her poorly grandmother. In order to keep her entertained, Kirstine stuck a video into the VHS. It was the Reader's Digest tape. She saw me and thought that I looked interesting. She then gave the tape to the development people at work the following day. It was they, several years later, who invited me onto the show. I'll never underestimate the power of Reader's Digest again!

210

Meanwhile, I had to build on my lucky break, nail the situation and get myself on TV more regularly. For the next few months I canvassed the BBC Natural History Unit (or NHU) in Bristol, essentially asking for a presenting job. I saw The Urban Birder not as myself but as an unusual concept that I could sell, and to do that job I had to draw upon my media sales experiences and develop a thick skin. Luck seemed to favour me and I very quickly amassed an ensemble of people who leapt to my aid, some of whom are close friends and confidants, and others whom are quite influential in the television world. A lady dogwalker at The Scrubs whom I had originally only known by sight at the beginning of 2006 became very chatty with me by April. She too saw my *Springwatch* piece and one day when we bumped into each other on The Scrubs she proclaimed that I needed an agent. I didn't understand why I needed one but the following day we met again and she gave me three agents' telephone numbers. I went home and rang the first one on the list. It turned out to be Bill Oddie's agent whose first words to me were, 'I'm glad that you called me as I was going to call you. I would like to represent you.'

I now had an agent with big name clients. A public relations girl I knew advised me to develop a website. I had a strong idea of exactly what I wanted to achieve with the site and opted for what I considered to be a slick and minimal domain that went against the normal birding website rules.

It was at this time that I reconnected with my great friend Stephen Moss, a series producer and director at the NHU. I had originally met Stephen in 1993 while twitching an Ortolan Bunting, a spring overshoot from continental Europe that had found its way to Richmond Park in London of all places. Stephen went on to help launch the very successful natural history career of Bill Oddie and worked on many other popular TV series. As if that wasn't enough, he has also written a whole slew of books on his favourite subject, birds.

211

Stephen very kindly spared the time to talk to me and offer his honest advice. He didn't pull any punches, telling me not to hold my breath, as competition was ultra stiff, with there not being enough work to go around for the existing roster of presenters. He proved to be right, and had

I realised just how difficult a nut I was trying to crack I would possibly have given up there and then.

In television, one of the most important groups of people to get excited and on your side are the production companies. These are the guys that generally discover talent and come up with and develop the bulk of the TV programmes that we see on the box. Once they can see a presenter's uniqueness and have designed a programme idea that could work for them, they then try to get an audience with the key people in power, the television commissioners. It is they who ultimately say yay or nay to the ideas presented. Getting to see those guys is harder than getting an audience with the pope if you are an ordinary and unknown mortal like me. Most of the production companies I met liked me and understood that I came into the picture from a different angle, but I did not appreciate just how difficult it was and still is to get programmes commissioned, no matter how much of a no-brainer idea you might have thought you were pitching.

As I mentioned earlier, a thick skin is a prerequisite as nothing moves quickly in television, so I changed my game-plan to become more patient. Then, out of the blue, I got called in for a meeting at the BBC by Doug Carnegie, who was then the Editor working on the launch of a new daily magazine-format show, soon to be aired on primetime BBC One. He offered me the chance to appear on his new show

212

despite me having virtually no real experience. That show was the soon to be lauded *The One Show*. To this day I am eternally grateful to Doug for having the belief in me and giving me my first break on television despite being green. Those are the kinds of breaks that you need in life. Things started to gather pace in my fledgling broadcasting career with regular appearances on *The One Show* plus snippets on a variety of different shows both in the UK and abroad. In addition, I was getting spots on the radio, usually talking about birds for BBC London, and I was a regular monthly guest on Tessa Dunlop's *Friday Night Show* where we talked about everything bar wildlife.

I have had some great experiences whilst filming *The One Show* pieces, like being pooped upon from a great height by an angry Lesser Black-backed Gull in Gloucester, handling Little Egret chicks in Norfolk and releasing Corncrakes into the wild in the Nene Washes, Cambridgeshire, but my over-riding favourite memory did not even involve birds. As a member of *The One Show* family I was invited to participate in 'The One Show Olympics' in which presenters were pitted against each other in different sports. I was chosen to run the 100 metres against fellow wildlife presenter Mike Dilger and resident hairdresser Michael Douglas. I was told about it a week before the actual race, which was to be filmed near the BBC's White City headquarters in the Linford Christie Stadium that funnily enough adjoins

Wormwood Scrubs. Running is not my forte. I'm the type of guy that would chose to walk rather than run after a bus, even if it meant being late if I missed it. I needed to get training, so who better to consult than my alternative health guru, Collin Flapper.

When Collin agreed to train me I envisioned hours of tedious hard work. However, knowing that I was capable of running at great speed from the times that we played football together, he said that the race would be won at the starting line even before the gun went. I laughed. Collin then told me a fantastic story about how a big premiership game between Manchester United and Arsenal was won before kick-off in the tunnel when Roy Keane the United skipper totally psyched out his counterpart Patrick Vieira.

Instead of getting me into the gym and forcing me to sprint around circuits, his training methods were bizarre to say the least. He got me to think and believe with every sinew in my body that I was going to win the race. He showed me how to position my body when in sprint mode by running on the spot and, most importantly, on the day he got me to dress for the occasion. That, he said, would be the psychological masterstroke and would win me the contest. So the fateful day came and I walked into the stadium dressed in my shorts and a funked up Beatles t-shirt, and with a pair of binoculars around my neck. As I started to stretch my limbs I got incredulous looks from the crew and I was pretty certain that I registered nervousness in the faces of my opponents. We stood at the starting line and I took my binoculars off only after looking at an oddly-plumaged second-winter Herring Gull that flapped past. Fully concentrating, I stared at the finish line ahead. It seemed like five miles away. I remembered what Collin said: "You are Usain Bolt and imagine sprinting towards an Eskimo Curlew sitting on the finishing line." The starter's gun went and I launched forward. Time stood still and everything went into a blur. I was convinced that both the Dilge and Douglas were on my shoulders, so I powered on. At the halfway point I was beginning to tire. The finish line was getting no closer but something in my head kicked in and I tried giving one supreme final effort. Suddenly, I was running through the tape. Exhausted, I turned to see that my opponents were miles behind me. The training had worked, Mike Dilger has never lived it down and I owed Collin as much Guinness as he could drink!

I have been fortunate enough to have visited many different cities around the world and my experiences have reinforced what I had felt all along: that many city-dwellers and birders alike just do not know the avian delights to be found on their home ground. There were so many times when the birder guiding me around their city was unaware of some of the key sites to be explored, and of the birds and other wildlife that lived there. I too was surprised by the abundance of birds in many of the cities that I explored. The Syrian Woodpeckers and Golden Orioles will always stick with me when I think of Budapest, along with the singing River Warblers and Common Rosefinches in Helsinki.

Even thinking about unfairly derided places like Croydon in Surrey remind me of a host of wonderful memories, such as hearing a drumming Lesser Spotted Woodpecker in Selsden Woods or studying Nuthatches as they crept around the branches of trees in urban Park Hill. For me it is all about gaining an appreciation of the natural life that surrounds us in our urban bubble.

216

I think that cities can be amazing places for wildlife, and I strongly implore everyone living in a city to go out and discover an area that has not been studied before. It could be a park or it could be an area of unused land, but whatever it is go and discover. It does not even have to be a conventional piece of habitat. One Saturday morning in October 2009 I was at the receiving end of a bad challenge by an opposing striker during a stint in goal at football. After the game we got talking and I discovered that he was a cameraman and had just been filming from the roof of Tower 42, one of the capital's tallest buildings in London's Square Mile. He could not stop raving about the view and the gorgeous sunrise.

For several years I had been very interested in getting up on an inner city rooftop to observe migration. I was inspired by the guys in the '50s and '60s who used to stand on top of Primrose Hill watching autumn migrants flood through London. They discovered the existence of a well-established migration route that cut directly through the city, again proving that birds do not necessarily avoid urban environments.

I got in contact with the management at Tower 42 the following week, to ask if I could bring up a small crew from BirdGuides to film the Woodpigeon movement that usually occurred at that time of year. In previous years, many thousands of Woodpigeons, possibly originating from Scandinavia, would pass over London from north-east to south-west, perhaps finally ending up in Iberia. It was a remarkable sight from the ground, especially set against a bright blue sky when the birds twinkled in the sunlight. I thought it would be fantastic to see that mass movement from a great height and across the London skyline.

To my surprise the lovely people at Tower 42 agreed, although on the day we could only summon around 200 'woodies' through the grey murk. I immediately saw the potential of this remarkable vantage point and a subsequent meeting between the Tower 42 Management Team and myself resulted, to their eternal credit, in the formation of the Tower 42 Bird Study Group.

The group met once a week during the spring of 2010 for a total of nine sessions and recorded a fascinating collection

The 360-degree panorama from Tower 42 is ideal for scanning for passing raptors and other migrants.

of species including daily Peregrines and Sparrowhawks as well as central London scarcities such as Oystercatcher, Arctic Tern, Common Buzzard, Red Kite, Hobby and, most famously, a couple of Honey Buzzards. One of the Honeys even managed to crash into a West End office window, much to the alarm of the office workers within. Fortunately the bird was unharmed and eventually continued on its journey. The autumn sessions resulted in several Sandwich Terns and another Red Kite.

Although it can be extremely hard work at times, akin to seawatching but with the waves being replaced by rows of buildings, Tower 42 has quickly become synonymous with visible migration in London. The great thing is that anyone can do something like this anywhere in the world and you don't even have to be an expert. With time the expertise will come, as all you need is passion and a love for nature. And that's just it. As I've said before, you don't need to be an expert to appreciate nature. You don't even need to know the names of the creatures that you are looking at.

People often ask me about what I would like to do in the future. Well, I think that there's a fantastic opportunity to get kids interested in birds and nature, especially urban kids who think that their local areas are devoid of wildlife. I would love to present a TV series in which I show youngsters just how much nature is entwined into their lives and how interesting

Black-headed Gulls are are classic urban birds, making themselves at home in towns and cities across Europe and beyond.

it is. When I was a young boy, apart from the occasional snippet on *Blue Peter* or *Magpie*, there were virtually no programmes on television aimed at getting kids engaged with the natural world. You had to rely on grown-ups filling you in or by discovering things for yourself. Nowadays, kids have many distractions to keep them preoccupied and disconnected from nature. But nature is everywhere. It's a major part of our culture, featuring in songs, books, poems and films. I think that it is all about making birds and other wildlife relevant, contemporary and exciting.

Speaking to kids in their language and getting them

involved with wildlife through using the technology that they have grown up with, like mobile phones and digital cameras, might get them motivated. Explaining that their favourite football team may have an animal-related nickname, that they may live on a street or in a town named after an animal or that their pet hamster or budgie originally had wild ancestors could all be great and fascinating stories. Getting a seven-year-old to collect 10 different types of animal from their garden or local park as part of a game or getting them to photograph or video the creatures that they come across would be great fun and educational. With the rate of global warming accelerating it is even more important that youngsters learn about the natural environment and, crucially, about the importance of the conservation of both wildlife and its habitats. With that knowledge, slowing down rate of the Earth's destruction can become second nature for tomorrow's adults.

221

I am so happy that my path in life has led me to this point, and whether the psychic I saw was right or not, it is a path that has become a major mission in my life. I truly hope that if you have not tried it already, you will walk out of your front door, whether you are at home, at work, on the way to school, away on business, or on holiday, and spend some time exploring your local environment for birds. Even if you have walked past it a thousand times before, it is always worthwhile to spare a moment to try and notice a bit of wildlife. All of us, regardless of age, can enjoy nature. You just need to do one simple thing: look up.

OTHER BIRDING TITLES
BY NEW HOLLAND

Advanced Bird ID Handbook:
The Western Palearctic
Nils van Duivendijk. Award-winning
guide covering key features of every
important plumage of all 1,350 species
and subspecies that have ever occurred
in Britain, Europe, North Africa and
the Middle East.
£24.99   ISBN 978 1 78009 022 1

Atlas of Rare Birds
Dominic Couzens. Amazing tales
of 50 of the world's rarest birds,
illustrated with a series of stunning
photographs and colour maps.
Endorsed by BirdLife International
£24.99   ISBN 978 1 84773 693 2

The Birdman Abroad
Stuart Winter. An account of the
overseas escapades of Britain's best-
known birding journalist, from
showdowns with illegal bird-trappers
in Malta to heart-warming tales of
conservation in Africa.
£7.99   ISBN 978 1 84773 692 5

Bird Songs and Calls
Hannu Jännes and Owen Roberts.
Perfect for the dawn chorus. CD
containing the bird sounds of 96
common species and an accompanying
book explaining about the songs and
calls and illustrating each bird with
colour photos.
£9.99   ISBN 978 1 84773 779 3

Chris Packham's Back Garden
Nature Reserve
Chris Packham. A complete guide
explaining the best ways to attract

wildlife into your garden, and to
encourage it to stay there.
£12.99   ISBN 978 1 84773 698 7

Colouring Birds
Sally MacLarty. Ideal gift to help
develop a child's interest in birds.
Features 40 species outlines – including
such favourites as Robin, Blue Tit,
Chaffinch and Green Woodpecker –
and a colour section depicting the
birds as they appear in life.
£2.99   ISBN 978 184773 526 3

The Complete Garden Bird Book
Mark Golley and Stephen Moss.
Best-seller which explains how to
attract birds to your garden and to
identify them once they are there.
Includes more than 500 colour
artworks of 70 species. Endorsed
by The Wildlife Trusts.
£9.99   ISBN 978 1 84773 980 3

Common Garden Bird Calls
Hannu Jännes and Owen Roberts.
Invaluable book and CD featuring the
songs and calls of 60 species likely to
be encountered in gardens and parks.
Each is illustrated with at least one
photo and a distribution map.
£6.99   ISBN 978 1 84773 517 1

Creative Bird Photography
Bill Coster. Illustrated with Bill Coster's
inspirational images. An indispensable
guide to all aspects of the subject,
covering bird portraits, activities such
as flight and courtship, and taking
'mood' shots at dawn and dusk.
£19.99   ISBN 978 1 84773 509 6

223

*A Field Guide to the
Birds of South-East Asia*
Craig Robson. New flexi-cover edition
of the region's only comprehensive field
guide. Fully illustrated in colour. Covers
all 1,300 species from Thailand, Laos,
Vietnam, Cambodia, Singapore,
Peninsular Malaysia and Myanmar.
£24.99 ISBN 978 1 78009 049 8

The Garden Bird Year
Roy Beddard. Gives both birdwatchers
and gardeners insights into how to
attract resident and migrant birds
to the garden, and how to manage
this precious space as a vital resource
for wildlife.
£9.99 ISBN 978 184773 503 4

The History of Ornithology
Valerie Chansigaud. The story of more
than two millennia of the study of birds.
Richly illustrated with numerous
artworks, photographs and diagrams,
including a detailed timeline of
ornithological events.
£17.99 ISBN 978 1 84773 433 4

Kingfisher
David Chandler. Beautifully illustrated
monograph detailing the life of this
spectacular species. Contains more
than 80 colour photographs. Also
available in the same series: *Barn Owl.*
£12.99 ISBN 978 1 84773 524 9

New Holland Concise Bird Guide
Endorsed by The Wildlife Trusts. An
ideal first field guide to British birds
for children or adults. Covers more

than 250 species in colour, comes in
durable plastic wallet to protect against
the elements, and also includes a fold-
out insert comparing similar species in
flight. Other titles available: *Butterflies
and Moths, Garden Wildlife, Insects,
Mushrooms, Seashore Wildlife, Trees* and
Wild Flowers.
£4.99 ISBN 978 1 84773 601 7

New Holland European Bird Guide
Peter H Barthel. The only truly
pocket-sized comprehensive field guide
to all the birds of Britain and Europe.
Features more than 1,700 beautiful
and accurate artworks of more than
500 species.
£10.99 ISBN 978 1 84773 110 4

Tales of a Tabloid Twitcher
Stuart Winter. The key birding
events and personalities, scandal and
gossip of the past two decades and
beyond seen through the eyes of a
birding journalist. A 'must-read'
book for all birdwatchers.
£7.99 ISBN 978 1 84773 693 2

Top Birding Sites of Europe
Dominic Couzens. An inspiration
for the travelling birder. Brings
together a selection of the best places
to go birdwatching in the continent.
Includes 175 photos, more than
30 locator maps and a CD of
bird sounds.
£22.99 ISBN 978 1 84773 767 0

**See www.newhollandpublishers.com
for further details and special offers**